Medical Error and Harm

Understanding, Prevention, and Control

Medical Error and Harm

Understanding, Prevention, and Control

Milos Jenicek

CRC Press
Taylor & Francis Group
Boca Raton   London   New York

CRC Press is an imprint of the
Taylor & Francis Group, an **informa** business

A PRODUCTIVITY PRESS BOOK

Productivity Press
Taylor & Francis Group
270 Madison Avenue
New York, NY 10016

© 2011 by Taylor and Francis Group, LLC
Productivity Press is an imprint of Taylor & Francis Group, an Informa business

No claim to original U.S. Government works

Printed in the United States of America on acid-free paper
10 9 8 7 6 5 4 3 2 1

International Standard Book Number: 978-1-4398-3694-1 (Hardback)

Library of Congress Cataloging-in-Publication Data

Jenicek, Milos, 1935-
    Medical error and harm : understanding, prevention, and control / Milos Jenicek.
       p. ; cm.
    Includes bibliographical references and index.
    ISBN 978-1-4398-3694-1 (hardcover : alk. paper)
    1. Medical errors. I. Title.
    [DNLM: 1. Medical Errors. 2. Medical Errors--prevention & control. 3. Safety
Management--methods. WB 100 J515m 2011]

    R729.8.J46 2011
    610.28'9--dc22   2010018836

Visit the Taylor & Francis Web site at
http://www.taylorandfrancis.com

and the Productivity Press Web site at
http://www.productivitypress.com

This book is not dedicated to all those who (think they) do not make errors; it is dedicated to the rest of us.

Contents

Author's Very Short Introduction: Minimizing Errors in Medicine

Thoughts to Think About

Cusjuvis hominis est errare, nullius nis insipientis in errore perseverare [To err is human, but to persevere in error is only the act of a fool].

—Marcus Tullius Cicero, 106–43 BC

Errare humanum est [To err is human].

—(Probably) Plutarch, 46–120 AD (Morals, cca 100 AD)

To err is human; to forgive, divine.

—Alexander Pope in 1711, 1688–1744

To err is human; to forgive is not company policy.

—Sign on a company bulletin board in Grand Rapids (undated)

It is human to err; and the only final and deadly error, among all errors, is denying that we have ever erred.

—G. K. Chesterton, 1874–1936

To err is human—but it feels divine.

—Mae West, 1893–1980 (attributed)

To err is human, but to really foul things up requires a computer.

—*Farmer's Almanac,* 1978

To err is human, but when the eraser wears out ahead of the pencil, you're overdoing it.

—J. Jenkins, 1843–1911

To err is human, to forgive is divine. While this may apply in general, forgiving, forgetting or ignoring errors in medicine isn't acceptable, since consequences could be disastrous.

Errors in everyday life lead to traffic accidents and their victims. Errors in industry injure workers and harm communities. Both situations are fertile grounds for errors that health professionals can make when caring for patients. Errors in medicine, then, are evidence that something has gone wrong in patient and community healthcare and that something has caused harm and should be prevented and corrected. This book is about such evidence and what physicians can and should do about it.

Examples of medical errors abound, such as unjustified exploratory and diagnostic procedures; foreseeable but unanticipated adverse effects of medical interventions or drugs; undesirable or incorrect surgical decisions and their outcomes; treatment unsupported by evidence of its effectiveness, efficiency, and efficacy.

All types of errors and their consequences, whether medical or otherwise, have multiple implications, such as their correction and prevention, legal pursuits, quests for repair and compensation, finding and implementing improvements, or evaluations of these initiatives and activities. Economic, social, physical, and mental health consequences are important for those committing errors and their victims. Modern life does not always simplify or eliminate error problems and challenges. It may, in fact, make them more frequent and sophisticated and more challenging to control.

At the social level, medical errors are matters taken to courts as topics of various litigations that should lead to corrections requested by plaintiffs and made by health professionals, their institutions, and working environments for the material and other compensation of victims by their perceived perpetrators.

Any medical error is a product of various "external" circumstances, including the environment, working conditions, and pressures; rapidly evolving technology; and managerial, administrative, or system functioning. These external factors only contribute to the essence (the internal factors) behind the medical error, namely, the physician's own faulty reasoning, logic, critical thinking, and decision making. "Internal factors" are about what happens "in our cranium," "in the craniums" of those with whom we are working, and "in the craniums" of those who created the working environment and tools of the healthcare situation at hand. They include our physiological and pathological attributes, attitudes, motor and sensory skills, as well as our responses to external factors. External factors are about what happens "outside the cranium."

Dealing with medical error is a learned experience like anything else. Is harm also caused by not teaching, learning, and understanding medical errors as faults in critical thinking? Understanding, preventing, and correcting such faults are the primary responsibility of every health professional.

Our message reflects the following essential theses:

- Medical error and harm, however interconnected, are not identical either on a theoretical level or on a practical level.
- The methodology of the study and management of medical error and harm is split between unique cases and multiple cases and events.
- Human (individual) and system error, however interconnected, are not identical; their understanding and control are methodologically complementary and more useful if handled separately.
- Uses of evidence regarding medical error and harm in dealing with them by way of argumentation, critical thinking, and informal logic is as important as producing the best evidence itself; both are necessarily interconnected.
- Medical error and harm are mass phenomena like disease and health; therefore, they must also be studied and controlled by epidemiological methods.
- However uneasy it may make some humanists, clinical care of individual patients and series of patients, health protection, and health promotion at both the individual and community levels also means "manufacturing health" (within the framework of stringent ethics and laws); medicine benefits from and must use experience with error and harm from "outside" sources such as industry, new technology development, transportation, business, economics, administration and management, finances, in addition to psychology, ergonomics, kinesiology, sociology, and biostatistics to humanely and effectively "produce" the best possible health of individuals and groups of individuals.

In this spirit, the book presents these elements, although not necessarily in the exact order indicated:

- A short history, concepts, methodologies of study and management, applications, and other experiences with error in general, across various domains of human activity, initiative, and endeavor.

- A semantic (definitions) and taxonomic (classification) overview and challenges in the medical error and harm domains.
- A methodology and experience overview in medicine and other health sciences.
- How studies, understanding, and decisions regarding errors should be carried out at various stages of clinical practice such as risk assessment in the patient, diagnosis, treatment, or prognosis.
- How to understand and deal with errors in the broader framework of medical care.
- How to understand and deal with errors in community medicine and public health.
- A discussion on whether dealing with errors is a learned experience and on why, where, and how we should implement and practice this kind of learning.
- The most important strategies and what we should do in the future pertaining to error in medicine.

A short executive summary in each chapter, expanding the customary conclusions of most current written messages in medicine, takes into account existing recommendations from business writing[1-3] and introduces each chapter, allowing the reader not only to better understand the message as a whole but also to decide if the chapter is applicable given his or her knowledge and experience. Conclusions in most of the chapters complete the executive summary and outline the following chapter. To make the message even more reader friendly, nominal and qualitatively assembled sets of items are bulleted where important.

This book is about errors in medicine such as those arising from various causes due to faulty information, thinking, reasoning, and communication leading to potentially serious mistakes in our understanding, decision making, and management of health problems. In fact, errors in reasoning, communication, and decision making underlie most of the other aforementioned reasons. Correct reasoning is, however, only one of many ways to avoid errors in what we do. Comments on harm that errors in reasoning and decision making often produce follow.

We also examine here the underlying problems of cognition of and critical thinking about health problems as a "person approach." Equally important is the "system approach," relevant in particular for surgical specialties and emergency medicine.

In human pathology, we learn about common underlying mechanisms and then individually about important health disorders and diseases and how to treat each one. Similarly, in dealing with errors attributed to critical thinking in medicine, we learn about paradigms, elements, and rules of critical thinking itself, and then we familiarize ourselves with their "pathology" (i.e., critical thinking disorders, flaws, and fallacies as diseases of reasoning) that ultimately produce medical error and its consequences. How can we understand, prevent, minimize, and otherwise control them?

Some first-time readers may ask where they should start and what they should read before tackling the human error problem to better understand the philosophy, objectives, and content of this book. Some notions and topics are repeated in various forms from one chapter to another. These repetitions are intentional to help experienced readers (who may choose to read selectively just one particular chapter) avoid a more laborious search for particular information elsewhere in the text. What might be the most useful for the overall domain?

We currently benefit from many first steps and initiatives in the development of methodologies, applications, and strategies and in the implementation and evaluation of the general and medical error and harm domain as seen in contributions from prominent and highly competent authors from around the world including in the United States, United Kingdom, Denmark, Italy, Australia, and Canada. It is now easier to move forward in this area.

The following "essential" references are only a few turning points and a partial list of background and complementary readings that have inspired this book. Read in chronological order, they appropriately reflect how our understanding of medical error is developing and maturing over time:

For human error in general: Jastrow,[4] Reason,[5] Cacciabue,[6] and Peters and Peters[7] with their numerous contributors summarize the human error problem over the past three generations.

For error in medicine: In general medicine, Fagerhaugh et al.,[8] Bogner et al.,[9] Kohn et al.,[10] and Vincent et al.,[11] have defined the error problem in the research and practice of medicine. In medical specialties, emergency medicine benefits the most so far from Croskerry et al.'s[12] coverage of patient safety and its determinants. In a similar spirit, Sharpe and Faden[13] offer a historical review of the transition from our vision and understanding of iatrogenesis and iatrogenic illness to our present grasp of the medical error and harm problem. Their monograph also illustrates well the still often confusing concepts and dimensions of error, harm, and iatrogenesis. It also suggests that the main problems lie in the fields of surgery, clinical pharmacology, hospital (nosocomial), and other infections and perhaps mental care.

For the review of strategies to manage error: *Medical Error*, edited by Rosenthal and Sutcliffe,[14] provides a 2002 update of the medical error situation as seen through past experience, by patients, healthcare providers (including evidence-based medicine), and system or human paradigms of medical error and harm. Its list of medical error and harm-related Web sites to know and use is also helpful. In addition, *Medical Mishaps*[15] compares **medical error experience in various parts of the world**.

For logic, epidemiology, and critical thinking–related methodological in medicine and health sciences: To set the medical error problem in the even broader context and longer tradition of more general principles, techniques, and methods, it is worthwhile to consult the current edition of the *Dictionary of Epidemiology*.[16] See Jenicek's *Foundations of Evidence-Based Medicine*[17] for general epidemiological topics, and Jenicek and Hitchcock's[18] *Evidence-Based Practice: Logic and Critical Thinking in Medicine* for fundamentals of modern logic and critical thinking in medicine. Or see some of Jenicek's other references for additional teaching and learning tools and handling of fallacies in medicine and as introductory readings to modern logic and critical thinking in our understanding and decision making in medical care.[19,20] They may prove useful as complementary reading. Thus, we are answering here the call for fundamental and clinical epidemiology and evidence-based medicine to complete and expand basic contributions of psychology, operational research, or computer sciences in the general human and medical error domains. Realizing that not all readers will be familiar with epidemiology, we have inserted some reminders of its principles and tools within related messages.

The domain of error in medicine and other health sciences is still in a state of flux, subject to multiple paradigms and not always convergent semantics, definitions, terminology, taxonomy, research methodologies, descriptions, search for causes, interventions, and their evaluations. It is not only relatively recent but also stems from many crucial and valuable contributions well beyond the health sciences, such as psychology, philosophy (logic/argumentation and critical thinking), ergonomics, operational research, health administration and evaluation, engineering, and computer sciences. These contributions cannot be omitted just for the sake of promoting the "purity" of medicine, but should be organically integrated as part of our "mainstream understanding" of the medical error challenge. Let us reflect on this diversity in the text with references to our preferred perspectives, as summarized again in the glossary at the end. This book stands side by side with longer-rooted concepts, terms, and approaches in clinical and fundamental epidemiology (risk, hazard, others), clinical pharmacology (adverse effects), and elsewhere across the entire medical field (e.g., outcome research, evidence-based strategies, knowledge translation).

An introductory example of iatrogenesis, error, and harm can illustrate this point. **Iatrogenesis** and **iatrogenic artefact** mean "adverse effects or complications caused by or resulting from medical treatment or advice" by any health professional in mainstream, alternative, or complementary medicines,[21] clinical and community care, and public physical, mental, or social health. Those effects are also termed iatrogenic illness (personal or individual perception: "I feel that I have a problem with my health"), iatrogenic disease (objective dimension: "Yes, this is a medical problem"), or sickness (perception by others: "He or she does not look well; he or she looks sick!"),[22,23] resulting from medical care. In this situation many questions arise immediately, namely:

- Is everything related to healthcare viewed as "iatrogenic?"
- Is everything avoidable, or is something inherent to what we are doing with patients (e.g., adverse effects of chemotherapy for cancer)?
- What exactly does "harm" mean?
- Do any erroneous understanding and action lead to harm?
- Are "error" and "harm" interchangeable notions? Are they different but complementary, yielding together even more powerful information for improvement in patient and community care?
- What does all this mean for our consideration of prevention, control, understanding, or other desire to do something useful to achieve the ideal of a most effective and harmless care?

Implications are important: economically, politically, legally, and professionally. For example, concerns about error and harm lead to the disproportionate practice of "defensive medicine" (nothing must be omitted), overmedication, overcare, number and extent of surgeries, their cost, legal implications, and social burdens. In many such instances, we and our patients expect less.

The chapters that follow will cover some additional elements and points to ponder in our quest to develop, advance, and practice the best medicine possible.

Our ultimate goal is to provide the highest level of quality of care and patient safety. Within this context, there are two crucial virtues and qualities to understand and adopt:

1. If we consider today **quality of care** as a level of performance and achievement assessed both by subjective value judgments and by more objective views of its structure, process, or impact,[14] then minimizing medical error and its impact is one of the cornerstones of quality-of-care assurance and management.
2. We currently view **patient safety** as the reduction and mitigation of unsafe acts within the healthcare system, as well as through the use of the best practices, shown to lead to optimal patient outcomes.[12]

Freedom from medical error, its prevention, and control of medical error itself and its consequences are at the core of patient safety concepts and strategies to deal with them. Effective prevention, care, and cures benefit both their recipients and their providers in competent and compassionate environments, both inside and outside hospitals and medical offices.

Before we examine this more closely, we want to acknowledge the assistance of key contributors to this project. The author remains grateful to several colleagues, friends, and experts whose invaluable insights helped make this book possible:

Joan Gilmour, professor of law at the Osgoode School of Law, York University, Toronto, for her gracious and meticulous review of Chapter 8.

Nicole Kinney, chief executive officer of Linguamax Services Ltd., for her expert review of language and style.

Rand MacIvor, creative design supervisor, Media Production Services at McMaster University, for the development and production of all artwork.

The administrative staff members of the Department of Clinical Epidemiology & Biostatistics at McMaster University, managed by Noël Fraser, for all their support in this initiative.

Kristine Mednansky, senior editor, and Tara Nieuwesteeg, project editor, at the Taylor & Francis Group for making this book possible and delivering it with hard-to-match expertise into the reader's hands.

Any remaining errors and flaws are imputable directly and solely to the author.

References

1. Anon (Oregon State University). *The Executive Summary.* 2 pages at http://oregonstate.edu/dept/eli/buswrite/Executive_Summary.html, retrieved December 3, 2008.
2. eHow Business Editor. *How to Write an Executive Summary.* 2 pages at http://www.ehow.com/how_16566_write-executive-summary.html, retrieved December 3, 2008.
3. Berry T. *Writing an Executive Summary.* 2 pages at http://articles.bplans/com/index.php/business-articles/writing-a-business-plan/writing-an-..., retrieved December 3, 2008.
4. *The Study of Human Error.* Edited by J Jastrow. New York: D. Appleton-Century Company, 1936.
5. Reason J. *Human Error.* Cambridge, UK: Cambridge University Press, 1990.
6. Cacciabue PC. *Guide to Applying Human Factors Methods. Human Error and Accident Management in Safety Critical Systems.* London: Springer Verlag, 2004.

7. Peters GA, Peters BJ. *Human Error. Causes and Control.* Boca Raton, FL: CRC/ Taylor & Francis Group, 2006.
8. Fagerhaugh SY, Strauss A, Suczek B, Wiener CL. *Hazards in Hospital Care. Ensuring Patient Safety.* San Francisco: Jossey-Bass Publishers (A joint publication on The Jossey-Bass Health Series and The Jossey-Bass Social and Behavioral Science Series), 1987.
9. *Human Error in Medicine.* Edited by MS Bogner. Hillsdale, NJ: Lawrence Erlbaum Associates, Publishers, 1994.
10. *To Err Is Human. Building a Safer Health System.* Edited by LT Kohn, JM Corrigan, MS Donaldson. Washington, DC: National Academy Press, 2000.
11. *Clinical Risk Management. Enhancing Patient Safety.* Second edition. Edited by C Vincent. London: BMJ Books, 2001. (First edition, 1995)
12. Croskerry P, Cosby KS, Schenkel SM, Wears RL. *Patient Safety in Emergency Medicine.* Philadelphia: Wolters Kluwer | Lippincott Williams & Wilkins, 2009.
13. Sharpe VA, Faden AI. *Medical Harm: Historical, Conceptual, and Ethical Dimensions of Iatrogenic Illness.* Cambridge, UK: Cambridge University Press, 1998.
14. *Medical Error. What Do We Know? What Do We Do?* Edited by MM Rosenthal, KM Sutcliffe. San Francisco, CA: Jossey-Bass, A Wiley Company, 2002.
15. *Medical Mishaps: Pieces of the Puzzle.* Edited by MM Rosenthal, L Mulcahy, S Lloyd-Bostock. Buckingham, UK: Open University Press, 1999.
16. *A Dictionary of Epidemiology.* Fifth edition. Edited by M Porta, S Greenland, Oxford, UK: Oxford University Press, 2008.
17. Jenicek M. *Foundations of Evidence-Based Medicine.* Boca Raton, FL: Parthenon Publishing Group, 2003.
18. Jenicek M, Hitchcock DL. *Evidence-Based Practice: Logic and Critical Thinking in Medicine.* Chicago: American Medical Association Press, 2005.
19. Jenicek M. *A Physician's Self-Paced Guide to Critical Thinking.* Chicago: American Medical Association Press, 2006.
20. Jenicek M. *Fallacy-Free Reasoning in Medicine: Improving Communication in Research and Practice.* Chicago: American Medical Association Press, 2009.
21. Wikipedia, the free encyclopedia. *Iatrogenesis.* 6 pages at http://en.wikipedia.org/wiki/Iatrogenesis, retrieved June 27, 2009.

Chapter 1

Putting Medical Error in Context: Minimizing Errors in Medicine— Beyond the "Oops!" Factor

Executive Summary

Making errors, understanding their causes and occurrences, and preventing them will always be an integral part of medicine, however regrettable this may be. In fact, error consequences may be disastrous for patients, healthcare workers, and entire communities. This chapter provides physicians and other healthcare providers with ways to address the medical error problem by offering a range of concepts, methods, techniques, and strategies grouped under the term *lathology*.

The medical error domain has many stakeholders including patients, physicians, other health professionals, legal specialists involved in litigation, health economists, sociologists, psychologists, ergonomists, and social workers. Patient safety as a whole may be considered synonymous with the absence of medical and other healthcare error in the practice and research not only of medicine but also of any health-related domain.

Current dealings with error in manufacturing, in new technology development, and its uses, and in transportation benefit from major contributions and developments brought about by many specialists working mainly in nonmedical fields. Health professionals are currently adding a new dimension to the increasingly integrated world of medical lathology.

Medical errors not only occur sporadically but may also be epidemic, endemic, and even pandemic in nature. Logically, clinical and field epidemiology are gradually focusing more on the search for medical error causes, investigation of their occurrence, and effectiveness of corrective programs and interventions. Their involvement in lathology is growing, as are the modern critical thinking argumentation and informal logic underlying medical reasoning, decision making, and epidemiological contributions to them.

A wide array of medical error causes such as inadequate training; failures of traditional and new medical technologies in their development and uses; physiological, psychological, and environmental influences; data and information management; execution deficiencies; health system functioning failures; communication breakdowns; rule-based errors; and errors in reasoning and decision making will be examined to varying degrees in subsequent chapters.

The notion of medical error is separate from medical harm. Medical error does not always lead to harm. Having said this, medical error and medical harm have specific causes that are sometimes synonymous and sometimes different. The study of both is crucial to improve patient safety.

Thoughts to Think About

Medical statesmanship cannot thrive only on scientific knowledge, because exact science cannot encompass human factors involved in health and disease. Knowledge and power may arise from dreams as well as from facts and logic.

—René J. Dubos, Mirage of Health (ch 6), 1959

Most men die of their remedies, not of their diseases.

—Molière, 1622–1673

Mistake, error, is the discipline through which we advance.

—William Ellery Channing, 1830 (1780–1842)

In medicine, there are many Popes, but none faultless.

—Karl H. Bauer, 1890–1978

If you shut the door to all errors, Truth will be shut out.

—Rabindranath Tagore, 1928

We have our faults and our virtues; we meet with failures and achieve success. many of our faults are entirely unavoidable, and arise from the fact that medicine is not an exact science … some things are quite impossible, and our work is carried out upon a jar, a chemical mixture in a retort, or a wooden Indian from the front of a cigar store.

—J. Chalmers Da Costa, 1863–1933

Every hospital should have a plaque in the physicians' and students' entrances: "there are some patients who we cannot help; there are none who we cannot harm."

—Arthur L. Bloomfield, 1888–1962

The pilot is by circumstances allowed only one serious mistake, while the surgeon may commit many and not even recognize his own errors as such.

—John S. Lockwood, 1907–1950

Don't make the wrong mistakes!

—Lawrence ("Yogi") Berra (1925–)

Not making errors in the practice of medicine is an essential attribute of patient safety, quality of clinical care, and community well-being. In health research, it is not only a prerequisite for academic and institutional (organization) promotion and honor, but it also carries the risk that erroneous research results will be perpetuated and magnified by the decisions of even the best-intended users, ending up eventually at courts of law. Errors in practice and research in medicine cannot be completely eliminated and avoided. Understanding their nature and nurture helps minimize them as much as possible. Despite the harm they cause, they are important lessons on how to improve what we know and do.

Introductory Comments: Errors as Part of Advances in Medicine

Medicine in clinical and community domains advances in many ways, including the following:

- Spectacular results in fundamental areas like stem cell research, medical genetics, or molecular explorations.
- Production, evaluation, and uses of the best evidence in basic and clinical pharmacology, surgical disciplines, and other clinical care covering all age groups.
- Continuous improvement of research, reasoning, critical thinking, and decision-making methodologies in all domains.
- Ever improving and refocusing medical education.
- Expanding structure, process, and impact of medical care evaluation including knowledge translation.
- Development of new technologies including their ethical context.
- Better focus, attention, and actions to understand, prevent, and control human and system errors in clinical care, community medicine, and public health within all the aforementioned domains as well as in the expanding experience resulting from their correction.

We often forget that learning from our own errors and correcting them is an extremely powerful educational and learning tool (if done right) and that our patients benefit immensely from otherwise unfortunate errors committed in the past. This is perhaps the greatest advantage of increasing the attention we give to the medical error domain.

Errors in medicine, so dreaded both by physicians and their patients, are undoubtedly more than warning evidence that something is wrong, causes harm, and should be prevented and corrected. This book is about such evidence and what physicians can and should do about it.

How to View Medical Errors Today

Medical errors happen in like risk assessment, diagnosis, treatment, prognosis, and related decisions. They occur also, sometimes endemically, in research and the practice of clinical, family, and community medicine or public health. Sometimes rare or not, expected or not, explained or not, they are an important part of the general error problem across various human endeavors. While most of the effort in medicine is focused on good evidence of beneficial actions and their results, uses, and effects, it should be noted that "bad" events such as medical errors, good evidence about them, as well as their control require equal attention, understanding, control, and prevention. The opposite would be contrary to medical ethics.

Medical errors also play an infrequently mentioned double role from the point of view of cause–effect relationships. On one hand, medical errors are caused by something. Methodologically speaking, they are dependent variables, consequences of something. We need to know their causes, to prevent and correct them. On the other hand, medical errors cause harm such as death or injury; errors here are the causes of harm and serve as independent variables in the association with their consequences. Harm itself may lead to a cascade of other consequences. Both cases and directions are relevant in our quest to provide quality care.

Medical errors belong to a larger family of errors across various domains such as errors in the development and use of new technologies, ergonomics, administration, management, politics, and economics. Experience in all these fields, acquired over the past three generations, is already partially applied in medicine. However, the specifics of medicine require additional attention to human and other factors affecting both care providers and their patients or health communities in the setting and context of their practice. Errors occur not only in fundamental, clinical, and community health research and evaluation but also in directly disturbing situations in daily practice and care. They also take place in knowledge translation (i.e., on the way from evidence producers to evidence users) and in beneficial or noxious consequences of evidence uses or nonuses.

Any medical error is then a product of various circumstances including the environment; working conditions and pressures; rapidly evolving technology, managerial, administrative, or system functioning; and other "external" factors. These external factors contribute only to the essence (i.e., the "internal factors") behind the medical error, namely, the physician's own faulty reasoning, logic, critical thinking, decision making, and his or her sensory-motor performance. This book mainly focuses on external factors where experience is still not as rich as it is for error in general. It also highlights internal factor circumstances where faulty reasoning and decisions occur. An interface between the former and the latter is the reality of our day-to-day life.

In human pathology, we learn about common underlying mechanisms and then about each individual important health disorder and disease and how to treat them. Similarly, in dealing with errors attributed to critical thinking in medicine, we learn about paradigms, elements, and rules of critical thinking itself, and then we familiarize ourselves with their "pathology" (i.e., their own disorders, flaws, and fallacies) as "diseases of reasoning" that ultimately lead to and produce medical error and its consequences. Without such learning and experience, how can we prevent and otherwise minimize medical errors?

Dealing with medical error is a learnt experience like anything else. Is harm also caused by not teaching, learning, and understanding medical errors as faults in critical thinking?

What Is Covered in This Book

As already mentioned in our short introduction to this book and the study of human error in general, we intend to expand this experience into the health sciences, clinical practice, and research and community medicine. Various steps of the medical cognitive process, argumentation, and decision making common to almost all activities will then be applied to all stages of clinical work such as assessing risk in the patient, making diagnosis, prescribing a treatment plan, making prognosis, and evaluating what was achieved and what was not or went wrong in this ever repetitive process. We will also review possible preventive measures against error and ways to control the error problem. Because errors in reasoning, communication, and decision making may occur—and they really do—in any of the aforementioned steps and activities, they are more than worthy of discussion in the chapters that follow.

Understanding and doing something about medical error also requires some basic knowledge of "shoe-leather-," research-, and clinical epidemiologies, modern informal logic, and critical thinking as outlined in some other writings. Readers are encouraged to search for any additional information beyond this book that they might need.

Just as our previous book was about fallacy-free reasoning in medicine,[1] this book is about errors and harm in medicine such as those arising for various reasons and causes from faulty information, thinking, reasoning, and communication in conjunction with other related error causes leading to potentially serious mistakes in our understanding, decision making, and management of health problems.

Considering the Medical Error Problem in Light of Recent Experience

From our recent discussion of the error problem in the fallacy domain of medicine,[1] let us now attempt to link the error problem to flawless reasoning and decision making.

Medical errors occurred too often in the past, occur today, and will unfortunately occur in the future. We must learn to live with them and do the best we can given the evolving circumstances of medical practice and research.

Several major medical periodicals[2-5] stress the urgency and magnitude of the problem in medicine[2-8] and surgery as well.[9-12] Some leading newspapers and magazines[13-16] and monographs[17] try to explain the challenge (especially diagnosis)[17] to a more general readership. The *American Iatrogenic Association*[18-20] as well as the *Level1Diet* initiative (709 references on care error so far)[21] offer a selection of major reports and articles dealing with the problem of medical errors. The University of Toronto's *Medicalerror* Web site,[22] and its numerous quality improvement links and lists, extends and updates reference readings. Leading national and international institutions initiate and further develop directions and strategies to handle the problem of errors in medicine and surgery.[23-27] The entire "medical error prevention and control" movement is gaining clearer objectives and focus and is becoming better structured and organized.

In a broader perspective of errors in medicine, we face the general problem of medical errors as the difference between actual behavior or measurement and the norms of expectations for the behavior or measurement.[28] More specifically for medicine, we face the problem of failures in planned action to be completed as indicated (error of execution) or use of an incorrect plan to achieve an aim (error of planning); the accumulation of errors results in accidents.[23] An error may then be an act of commission or an act of omission.[29] For example, in surgery, an error is more than tying a bad knot, a poorly executed suture. Many medical errors are, in a broader sense, clinical errors that can be made by other health professionals or when working together.

In more general terms, Bruce Bagley, past president of the American Association of Family Physicians, is perhaps correct in saying that "...a medical error is anything that happened in my office that shouldn't have happened and that I absolutely do not want to happen again."[24]

Errors in medicine are imputable to several reasons:

■ Inadequate training (knowledge, attitudes, skills).
■ Failure of medical technologies ("the machine is poorly designed or broken").
■ Inappropriate uses of medical technologies ("the tool is used where, when, and in whom it should not be").
■ Physiological and psychological factors such as the physician's, other health professional's, or patient's condition and disposition like fatigue or stress.
■ Data and information recording, processing, and retrieval caused by information technology and its uses (information technology inadequacy and failure).
■ Deficient skills in execution (motion or sensory activities as based on past experience).
■ Taxonomical errors due to causes of errors (classification of faulty activities due to poorly explained or used etiology).

- System (health services functioning) failures ("triage and subsequent emergency care does not work as it should").
- Communication errors and breakdowns.
- Rule-based errors (guidelines, user guides not followed).
- Errors in reasoning and decisions about health problems.

The chapters that follow are mainly about the last item in the list. In fact, errors in reasoning, communication, and decision making underlie most of all the other previously mentioned reasons. Fallacy-free reasoning is, however, just one of many ways to avoid errors in what we do.

Making a medical mistake is not necessarily malpractice with all its legal and financial consequences, but it may be so.[30] It may also cause (or not) some kind of harm. This book is about errors in reasoning and decisions in medicine beyond the malpractice problem.

Across the literature, medical errors are studied and evaluated in two ways that are not always clearly specified. One approach is to investigate causes of medical errors (i.e., errors are consequences or dependent variables). In the other approach, medical errors are linked as causes to harm (i.e., errors are causes or independent variables). Current taxonomies of medical errors do not always specify this possible double role of errors.

Medical errors are not limited to diagnosis or treatment decisions. They may occur at any stage of medical work: assessing the risk of disease, understanding its causes, and effectiveness of intervention to prevent or cure or otherwise control a health problem or its prognosis at an individual or community level.

Medical errors may also be studied through quantitative methods such as biostatistics or computer science, through methods adopted from other domains such as aviation,[31] or through qualitative methods.[32] With the recognized place of humanities in medicine,[33] the door opens to informal logic and critical thinking[34] (a natural companion to evidence-based medicine and clinical epidemiology[35]) as guardians against medical errors.

In this book, one of many trying to contribute to the minimization of medical errors, we are strongly interested (compared with others) in the underlying problems of cognition of and critical thinking about health problems[2,34,36,37] as a *person approach*. An equally important approach is the *system approach*, relevant in particular for surgical specialties and emergency medicine. Medical informatics and computer science not only contribute to a better understanding and management of medical errors through detection and surveillance systems or understanding of systems themselves[38]; they also enrich the current taxonomy of medical errors.[39-41] Communication breakdowns and errors are in focus too.[42] Interpretation and evaluation slips and mistakes, ambiguous information, or heuristic situations are categories used by informaticians[40,41] reflecting, under different names, errors in reasoning, argumentation, and decision making as outlined and illustrated in the chapters that follow.

Links between evidence-based medicine and critical thinking are being established,[34,35,43] and cognitive psychology is even being proposed as one more basic science for medicine.[44]

Medical Error and Patient Safety

Croskerry et al.[45] defined *patient safety* as "the reduction and mitigation of unsafe acts within the healthcare system, as well as through the use of best practices, shown to lead to optimal patient outcomes." The World Health Organization (WHO) Alliance for Patient Safety sees it as "freedom from unnecessary harm or potential harm associated with healthcare;" *harm* implies "impairment of structure or function of the body and/or any deleterious effect arising therefrom."[46]

Hence, patient safety is compromised by harm. Harm itself is often due to error. Error in this context is a failure to carry out a planned action (elements in healthcare in our case) as intended or the application of an incorrect plan and may manifest itself by doing the wrong thing (an error of commission) or by failing to do the right thing (an error of omission), at either the planning or execution phase.[46] Avoiding medical error is one of the cornerstones of patient safety.

National and international institutions, healthcare professionals, professional associations, governments, healthcare, and the legal system are paying, each in its own way and for its own purposes, bringing increasing attention to patient safety and consequently to medical errors.

Glossaries, taxonomies, original research, and data and information gathering through various surveillance and other systems are growing in numbers. The original methodology from the error problem in society and its components is being adapted and expanded for healthcare and all health professions involved.

Recognizing the need for teaching and learning to ensure patient safety in the aforementioned spirit and dimension, WHO now offers a patient safety curriculum for medical schools including some teaching summaries and audiovisual tools for the undergraduate level.[47]

Besides the political and strategic support and direction of international bodies like the WHO World Alliance for Patient Safety or the Linnaeus-PC Collaboration, several national agencies were created to coordinate initiatives in the medical error and patient safety domains, to standardize and expand research methodology, and to offer harmonized strategies, including the Canadian Patient Safety Institute, the Institute for Safe Medication Practices Canada, the National Patient Safety Agency (UK), the U.S. Agency for Healthcare Research and Quality (AHRQ), the Joint Commission (formerly the Joint Commission on

Accreditation of Healthcare Organizations, JCAHO), the Institute of Medicine's Committee on Quality of Healthcare in America, and the U.S. Department of Veteran Affairs' National Center for Patient Safety.

Based on the history of the worldwide expansion of attention to medical error and patient safety, Hofoss and Delikas[48] propose directions for performing patient safety research:

■ Specific investigation of adverse event cases.
■ Delivery system reviews.
■ Study of organizational culture.
■ Patient safety culture.

In addition to such directions, let us add refinements in medical error study methodology an enhancement of its tools beyond the crucial contributions of psychology, operational research, information technology applications, uses, and advancement. Epidemiology and medical specialties themselves represent some still underused approaches in the core of the medical problem challenge.

We are entering a domain with its own specific identity that we may call **lathology**, the domain of study and control of medical error and harm. This will be discussed in greater detail in Chapter 2.

Subsequent chapters should then be, we hope, on the right track.

How This Book Might Contribute to the Present State of Human Error Experience and Patient Safety

The chapters that follow will not provide answers to all questions about how to make the best decisions in medicine and performing clinical and extraclinical acts well. However, they should at least support your own justifications for finding answers to your questions. It's all about making sense. Fallacy-free reasoning and decision making is as important for limiting the frequency and severity of medical errors as a safe, effective, and properly used new medical technology. The critical thinking approach is only one among many helping us understand and better solve the problem of medical errors. Nevertheless, it is important.

Errors result not only from an individual person's failures but also and more often from failures of systems, various elements of physical and human environments, and their interconnections.[49,50] Extensive research on this topic must still be done.[51]

Both medical error and harm are entities that should be examined separately and connections between them should be sought. In an extreme view, patients do not care if some medical error was committed, but if some harm to their health does occur, they want to know if such harm is related to medical error and how it should be explained, corrected, prevented, and compensated.

In human pathology, we learn about common underlying mechanisms and then about each individual important health disorder and disease and how to treat them. Similarly, let us stress again that, in dealing with errors attributed to critical thinking in medicine, we learn about paradigms, elements, and rules of critical thinking itself, and then we familiarize ourselves with their own "pathology" (i.e., "critical thinking disorders, flaws, and fallacies as diseases of reasoning"). The long journey described and explained in the following pages is worth taking and perhaps necessary to reach our *primum non nocere* ideal.

References

1. Preface. Contributing to reducing errors in medicine. Pp. xix–xxiv in: Jenicek M. *Fallacy-Free Reasoning in Medicine. Improving Communication and Decision Making in Research and Practice.* Chicago: American Medical Association Press, 2009.
2. Leape LL. Error in medicine. *JAMA*, 1994; **272**:1851–7.
3. Berwick DM, Leape LL. Reducing errors in medicine. *BMJ*, 1999; **319**:136–7.
4. Bates DW, Gawande AA. Error in medicine: What have we learned? *Ann Intern Med*, 2000; **132**:763–7.
5. Handler JA, Gillam M, Sanders AB, Klasco R. Defining, identifying and measuring error in emergency medicine. *Acad Emerg Med*, 2000; **7**:1183–8.
6. Reinertsen JL. Let's talk about error. *BMJ*, 2000; **320**:730.
7. Wu AW. Medical error: the second victim. *BMJ*, 2000; **320**:726–7.
8. Reason J. Human error: models and management. *BMJ*, 2000; **320**:768–70.
9. Cuchieri A. Surgical errors and their prevention. *Surg Endosc*, 2004 (Online publication 2005); **19**:1013.
10. Satava RM. The nature of surgical error. A cautionary tale and the call to reason. *Surg Endosc*, 2005; **19**:1014–6.
11. Dankelman J, Grimbergen CA. Systems approach to reduce errors in surgery. *Surg Endosc*, 2005; **19**:1017–21.
12. Cuschieri A. Reducing errors in the operating room. Surgical proficiency and quality assurance of execution. *Surg Endosc*, 2005; **19**:1022–17.
13. Gawande A. When doctors make mistakes. *The New Yorker*, February 1, 1999.
14. Leonhardt D. Why doctors often get it wrong. *The New York Times*, February 22, 2006.
15. Gorman C. Where doctors go wrong. *Time*, March 7, 2007. Available at http://www.time.com/time/magazine/article/0.9171.1599718.00.html, retrieved December 16, 2007.
16. Groopman J. The mistakes doctors make. Errors in thinking too often lead to wrong diagnoses. *The Boston Globe*, March 19, 2000.
17. Groopman J. *How Doctors Think*. Boston: Houghton Mifflin Company, 2007.

18. American Iatrogenic Association Library. *Medical Error. Major Reports.* Available at http://www.iatrogenic.org/library/mederrorlib3.html, retrieved December 16, 2007.

19. American Iatrogenic Association Library. *Medical Error. 1988–2000.* Available at http://www.iatrogenic.orglibrary/mederrorlib/html, retrieved December 16, 2007.

20. American Iatrogenic Association Library. *Medical Error. 2001–Present.* Available at http://www.iatrogenic.org/library/mederrorlob2.html, retrieved December 16, 2007.

21. *Level1DietTM. Care Errors. Health Information Search Results. Matching Summaries of Recent Peer Reviewed Scientific Research Care Reports.* Available at http://www.level1diet.com/care%20errors_q, retrieved December 26, 2007.

22. MacDonald C, Howard F. *Medical Error. Ethical Aspects of Clinical Error and Patient Safety.* A University of Toronto website and subsites (links, quality, books). Available at http://www.medicalerrors.ca, retrieved December 15, 2007.

23. *To Err Is Human: Building a Safer Health System.* Edited by LT Kohn, JM Corrigan, MS Donaldson. Committee on Quality of Health Care in America. Washington, DC: National Academy Press (Institute of Medicine), 1999 (2000).

24. California Academy of Family Physicians. *Diagnosing and Treating Medical Errors in Family Practice.* Available at http://familydocs.org.assets/Publications/Monographs/Monograph/MedErrors.pdf, retrieved December 17, 2007.

25. World Health Organization. *First International Consultation on Improving the Safety of Surgical Care.* January 11–12, 2007, Geneva Switzerland. Available at http://www.who.int/patientsafety/events07/11_01_2007/en/index.html, retrieved December 19, 2007.

26. World Health Organization. *The Second Global Patient Safety Challenge: Safe Surgery Saves Lives,* background paper and draft for discussion, First International Consultation Meeting, December 12, 2006, Available at http://www.who.int/patientsafety/events07/11_01_2007/en/index.html, retrieved December 19, 2007.

27. World Health Organization. *Meeting Summary.* First Consultation, *Safe Surgery saves Lives,* WHO Headquarters, Geneva, January 11–12, 2007, Available at http://www.who.int/patientsafety/events07/11_01_2007/en/index.html, retrieved December 19, 2007.

28. Wikipedia, the Free Encyclopedia. *Error.* Available at http://en.wikipedia.org/wiki/Error, retrieved December 12, 2007.

29. Wu AW, for the Committee on Identifying and Preventing Medication Errors. Preventing Medication Errors. Presentation at the 14th Annual National Symposium on Patient Compliance. (Healthcare Compliance Packaging Council and Institute of Medicine of the National Academies), Baltimore, MD, May 16, 2007.

30. Holder AR. Medical Errors. *Hematology 2005*; **2005**(1):503–6.

31. Catchpole KR, Giddings AEB, Hirst G, Dale T, Peek GJ, de Leval MR. A method for measuring threats and errors in surgery. *Cogn Tech Work*, electronic version, DOI 10.1007/s1011-007-0093-9, Springerlink date Friday, July 27, 2007. 10 pages at http://www.springerlink.com/content/n476067uq62182g8/, retrieved December 12, 2007 and January 3, 2008.

32. Kuzel AJ, Woolf SH, Engel JD, Gilchrist VJ, Frankel RM, LaVeist TA, et al. Making the case for a qualitative study of medical errors in primary care. *Qual Health Res,* 2003; **13**(6):743–80.

33. Cassel EJ. *The Place of the Humanities in Medicine.* Hastings-on-Hudson, NY: Institute of Society, Ethics and Life sciences, The Hastings Center, 1984.
34. Jenicek M, Hitchcock DL. *Evidence-Based Practice. Logic and Critical Thinking in Medicine.* Chicago: American Medical Association Press, 2005.
35. Jenicek M. *Foundations of Evidence-Based Medicine.* Boca Raton, FL: The Parthenon Publishing Group/CRC Press, 2003.
36. Croskerry P. The cognitive imperative: Thinking about how we think. *Acad Emerg Med,* 2000; **7**(11):1223–31.
37. Croskerry P. The importance of cognitive errors in diagnosis and strategies to minimize them. *Acad Med,* 2003; **78**(8):775–80.
38. Bates DW, Cohen M, Leape DL, Overhage MJ, Shabot MM, Sheridan T. Reducing the frequency of errors in medicine using information technology. *J Am Med Inform Assoc,* 2001; **8**:299–308.
39. Kopec D, Kabir MH, Reinharth D, Rothschild O, Castiglione JA. Human errors in medical practice: Systematic classification and reduction with automated information systems. *J Med Systems,* 2003; **27**:297–313.
40. Zhang J, Patel VL, Johnson TR, Shortliffe EH. Toward an action based taxonomy of human errors in medicine. *Proceedings of 24th Conference of Cognitive Science Society.* Available at http://acad88.sahs.uth.tmc.edu/research/publications/cogsci2002-taxonomy.pdf, retrieved December 17, 2007.
41. Zhang J, Patel VL, Johnson TR, Shortliffe EH. A cognitive taxonomy of medical errors. *J Biomed Informatics,* 2004; **37**:193–204.
42. Greenberg CC, Regenbogen SE, Studdert DM, Lipsitz SR, Rogers SO, Zinner MJ, et al. Patterns of communication breakdowns resulting in injury to surgical patients. *J Am Coll Surg,* 2007; **204**:533–40.
43. Kee F, Bickle I. Critical thinking and critical appraisal: the chicken and the egg? *QJM,* 2004; **97**(9):609–14.
44. Redelmeier DA, Ferris LE, Tu JV, Hux JE, Schull MJ. Problems for clinical judgment: introducing cognitive psychology as one more basic science. *CMAJ,* 2001; **164**(3):358–60.
45. Croskerry P, Cosby KS, Schenkel SM, Wears RL. *Patient Safety in Emergency Medicine.* Philadelphia: Wolters Kluwer | Lippincott Williams & Wilkins (Health), 2009.
46. World Alliance for Patient Safety. *The Conceptual Framework for the International Classification for Patient Safety. Version 1.0 for Use in Field Testing 2007 – 2008. (ICPS).* Geneva: World Health Organization, July 2007. Available at http://www.who.int/patientsafety/taxonomy/icps_download/en/print.html, retrieved February 4, 2009.
47. World Alliance for Patient Safety. *WHO Patient Safety Curriculum Guide for Medical Schools. First Edition Draft.* Publication WHO/ER/PSP/2008.13. Geneva: World Health Organization, 2008. Available at http://www.who.int/patientsafety/activities/technical/medical_curriculum_form/en/index...., retrieved February 4, 2009.
48. Hofoss D, Delikas E. Roadmap for patient safety research: approaches and roadforks. *Scand J Public Health,* 2008; **36**:812–7.
49. Vincent C. Understanding and responding to adverse events. *N Engl J Med,* 2003; **348**(11, March 11):1051–6.

50. Delbanco T, Bell SK. Guilty, afraid and alone—struggling with medical error. *N Engl J Med,* 2007; **357**(17, Oct 25):1682–3.
51. Altman DE, Clancy C, Blendon RJ. Improving patient safety—five years after the IOM report. *N Engl J Med,* 2004; **351**(20, Nov 11):2041–3.

Chapter 2

The Valued Legacy of Error and Harm in General: Error and Harm across General Human Experience in Nonmedical Domains— Welcome to Lathology

Executive Summary

Modern general human error experience dates back to the 1930s. Its development is still facing heterogeneous terminology and taxonomy challenges, with additional methodological requirements for human error occurrence studies, search for causes, as well as prevention and control methodology development, implementation and evaluation. Medical error, in addition to such difficulties, requires adaptations specific to the medical working environment, disease history, medical care, and related human behavior in the health context.

Medical error is most often viewed as a system error that includes environmental factors, technology, working processes, and interactions between those components. Alternatively, medical error is viewed as a fault due to an individual as final operator and decision maker at the endpoint of system happenings, taking into account human mental cognition processes and sensorial-motor activities. The third view, integrating both system and person paradigms, is yet undeveloped. Person-oriented models will increasingly include uses of flawless and fallacy-free reasoning, critical thinking, and decision making to explain medical error and to find relevant solutions.

Because most of the aforementioned processes and strategies allow either avoidance or detection of errors whose causes are already known, uses of epidemiology in medical error within the context of "disease of interest" study and management are worthy of wider use; early initiatives are promising. In fact, there is often no better way to identify "new, still unknown causes." Medical error must be studied both as a consequence of some causal exposures as well as a cause itself of harm and various other outcomes from an instant and prognostic view.

From this error problem in the nonmedical domain, Chapter 3 will examine human error and harm in the health sciences in both clinical and community medicine and public health settings.

Thoughts to Think About

All the errors of politics and in morals are founded upon philosophical mistakes, which, themselves, are connected with physical errors. There does not exist any religious system, or supernatural extravagance, which is not founded on an ignorance of the laws of nature.

—Marie Jean Nicolas de Caritat, Marquis de Condorcet, 1794

Reason and free inquiry are the only effective agents against error.... They are the natural enemies of error, and of error only.

—Thomas Jefferson, 1743–1826

A hallucination is a fact, not an error, what is erroneous is a judgment based upon it.

—Bertrand Russell, 1872–1970

The error of Louis XIV was that he thought human nature would always be the same. The result of his error was the French Revolution. It was an admirable result.

—Oscar Wilde, 1854–1900

Xerox is a copying device that can make rapid reproduction of human error possible.

—Thinkexist.com, 2008

The harm done is often difficult to repair.

—Yusuf Islam, formerly Cat Stevens, born Steven Demetre Georgiou, 1948

It may seem a strange principle to enunciate as the very first requirement in a hospital that it should do the sick no harm.

—Florence Nightingale, 1820–1910

As to diseases, make a habit of two things—to help or at least, to do no harm.

—Hippocrates in Epidemics, ca. 460 BC–ca. 370 BC

Why is medical error and harm so important to us? Because our own survival, notwithstanding our lifestyle and the rest of medical care, is at stake.

Introductory Comments

Errors in medicine are evidence that something goes wrong, often causes harm, and should be prevented and corrected. This book is about such evidence and what physicians and other health professionals can and should do about it. Initially, the principal subject of any error study was an *event*; the patient and his or her disease and cure came only later.

In this second chapter, let us look first at general experience with the error problem across various human activities and domains of endeavor before turning to the specifics of error in health sciences. The purpose of this chapter is not to criticize the general error experience. Instead it is an opportunity to review the current spectrum of experience. Nonetheless, some strengths and weaknesses of current experience are retained for consideration of what to improve in specific situations of medicine and other health sciences. Hence, where are we in the error domain today?

We make errors almost anywhere, anytime, anyplace: driving cars; flying planes; making lifestyle choices; operating industrial and agricultural machinery; electing politicians; designing, producing, and using new technologies; choosing those with whom we wish to work or live. Occasional "accidental" disasters like Chernobyl or Three Mile Island are direct consequences of human error in the running of an industrial complex. All these domains are often subject to rapid technological and other changes over time and from one generation to another, which explains the current diversity of our views of human error as outlined in this chapter. Medical and clinical errors will follow under a separate title. Their current view is largely based on broader human experience, well beyond the health domain: our current knowledge of the error problem is based on contributions of experts from the fields of ground transportation, aviation, information science, manufacturing, new technologies development, and military science among others.

How then can we define the domain of error and harm today? The following section answers this question.

A Brief History of Recent Human Error Experience

Modern experience with human error spans a little more than the past three generations. It is closely related to the late fallouts of the industrial revolution such as changes in working environment and conditions, interpersonal communication and relationships, further development of new technologies, transportation, and technology of communication among others—hospital and medical office environments included. Health sciences and its technologies closely followed industrial and technological revolutions, trends, and developments. In addition, modern critical thinking, reasoning, and problem solving significantly enhanced our understanding of the human error challenge.

In the 1930s, a group of authors led by psychologist Joseph Jastrow[1] published the story of human error from the perspective of various basic sciences, including astronomy, geography, physics, and chemistry but also closer to medicine, zoology, physiology, neurophysiology, anthropology psychology, sociology, medicine, and psychiatry themselves. Such experiences were then organized, conceptualized, and structured by a number of psychologists, information technology experts, operational researchers, mechanical engineers and others, before being imbedded in an increasing number of monographs like those of James Reason[2] or reviews by groups of experts,[3] shared with an ever growing readership. Needless to say, the problem of human error and its control is also increasingly relevant for the military (U.S. Air Force) and space exploration (NASA).

Informal conferences of specialists (Columbia Falls, Maine, July 7–9, 1980) and events institutionally sponsored by the North Atlantic Treaty Organization's (NATO's) Science Committee and the Rockefeller Foundation have taken place. Their participants contributed to further standardize views and concepts while also presenting some less heterogeneous terminology, trends, and objectives in research and in the practice of dealing with human error.[4]

In the study of human error, the term *errorology*[5] (or rather *error-logy*) was proposed as an embodiment of logical and psychological trends in the search for truth. As a replacement for this kind of tongue twister, in the spirit of what has already been discussed, let us consider instead from the Greek *lathos* (bug, error, fault, mistake, oversight) and *logos* (word, reckoning, recounting, gathering) the term **lathology** for "the rational principle that governs and develops the error universe." We will use this term and its derivatives like *lathologist* and *lathometrics* throughout this book. Just as biometrics or econometrics mean measurement and analysis of life and economy phenomena, *lathometrics* means counting the frequency of error, measuring its dimensions, as well as categorizing error and error-related phenomena.

The ideal of truth relies on the absence of errors. Lathology includes the correction of errors that prevail in regard to logical procedures and the relationships between their components. It also includes those attributable to the fallible "common sense."[2] Other errors are more technical in nature.

Definition of Human Error and Other Related Terms

Nonetheless, clearer general (transdisciplinary) definitions of human error came only later. For Reason,[6] these definitions encompass attention failures, memory lapses, unintended words and actions, recognition failures, inaccurate and blocked recall, errors of judgment, and reasoning errors. Intentional actions without prior intention, nonintentional or involuntary actions, unintended actions, intended actions, and mistakes (mistakes as planning failures and slips and lapses as execution failures) are embodiments of error.[2,6]

How then can we define human error in operational terms today? Can we classify it somehow based on a practical and pragmatic approach to human error, its understanding, control, and prevention? What are the strategies and directions to deal with human error stemming from the experience we acquired over time? In this chapter, let us propose some answers to such questions.

We may feel that error is a mistake of some kind. But are there more concrete formulations? In fact, there are several, depending on whom you ask: philosophers, statisticians, ergonomists, lawyers, engineers, and management or information experts. The persisting diversity is understandable because those definitions were established with different theoretical and practical objectives in different domains of activity. Table 2.1 summarizes some major definitions of error and selected other items related to it.[7-27] Part of the challenge of dealing with error today is the heterogeneity of definitions and what people understand "error" to be. In this table, we offer a collection of definitions in a sequence reflecting their prevalent presentation across the literature. The glossary at the end of this book is organized in alphabetical order.

Note about Heterogeneity of Terms

These types of similarities and dissimilarities between definitions of error and related phenomena largely define methodological and practical approaches in the error domain, making it practically impossible at least for now to present a uniform and unequivocal picture of the error problem, in our context. The glossary at the end of this book summarizes definitions we prefer. This book should be read bearing in mind the meanings in the glossary.

Different meanings of the previously mentioned terms are still a reality across the literature covering error in general and error in medicine. The reader should be aware of this fact before adopting definitions and their uses further in this text. The problem is compounded by numerous differences compared with epidemiological terminology and definitions, more familiar to an error uninitiated health professional. Wherever relevant, we will stress such differences and bridge the gap between error in general and health domains.

Note about Error versus Accident

We may note from definitions[7-27] in Table 2.1 that error and accidents are not the same thing. Error is seen mainly as a failure of mental, judgmental, and inferential processes resulting in incorrect decision making and ensuing actions. Error should be seen as failures that lead to (are potential causes of) accidents as previously defined.

All these definitions were proposed by proponents from different domains of expertise, be it psychology, philosophy (logic and critical thinking), biology, engineering, ergonomics, or operational research. Such a variety of definitions and their origins contribute to the difficulty in identifying more uniformly errors, counting their occurrence, explaining their causes, and laying grounds for their correction and prevention. Some definitions are mostly conceptual, and not all are operational enough. Moreover, so far they do not allow a proper universal categorization of errors and their more useful taxonomy; overlapping of many definitions is a reality.

There is no unified view of the relationship between errors and accidents. Errors may be considered causes of accidents or not. Dekker[28] proposes a

Table 2.1   Some Definitions of Error and Related Terms across Human Experience

Semantic
"A wandering about," "straying."[7,8]
Lexical
The condition of having incorrect or false knowledge.[9] • An act, assertion, or belief that unintentionally deviates from what is correct, right, or true.[9] • An act, an assertion, or a decision—especially one made in testing a hypothesis—that unintentionally deviates from what is correct, right, or true.[10]
Philosophical
• Missing the truth; a consequence of absence of knowledge (and an excess of will).[8] • Deviation from accuracy or correctness.[7]
Legal
An incorrect ruling by the judge, which may be: • *harmless* (it did not affect the outcome of the case or prejudice a substantial right of a party) • *plain* (so obviously prejudicial to substantial rights of a party that it amounts to an affront to the judicial system) • *reversible* (prejudicing the appellant in a way that could have affected the outcome of the trial; if properly objected, it requires modification or reversal of the judgment).[11]
Mathematics, Statistics, Experimental Science
• Difference between a true value and an estimate, or approximation, of that value.[12] • Difference between the mean of an entire population and the mean of the sample drawn from that population.[12] • A difference between a computed, estimated, or measured value and the true, specified, or theoretically correct value.[7]

Continued

Table 2.1   Some Definitions of Error and Related Terms across Human Experience (*Continued*)

• In experimental science: The difference between a measured value and the true value of a quality or attribute.[7]
Engineering
• A difference between a desired and actual performance or behavior of a system or object.[7]
Telecommunications
• A deviation from a correct value caused by a malfunction in a system or a functional unit.[7]
Computer Programming and Software Engineering
• An incorrect action or calculation performed by software. Also, incorrect actions on the part of a program.[7]
Biology
• A loss of perfect fidelity in the copying of information, either good or bad.[7]
Philately and Numismatics
• Printing or production mistake that differentiates from a normal specimen or from intended result.[7]
Aviation
• Series of actions and assessments that are systematically connected to people's tools and tasks and environment.[13] (N.B. They are not considered surprising brain bloopers.)
Linguistics
• Deviation from standard language norms in grammar, syntax, pronunciation, and punctuation.[7]
Transdisciplinary (Error in General)
• "All those occasions in which a planned sequence of mental and physical activities fails to achieve its intended outcome, and when these failures cannot be attributed to the intervention of some chance agency."[2]

Table 2.1 Some Definitions of Error and Related Terms across Human Experience (*Continued*)

• The failure of planned actions to achieve their desired ends without the intervention of some unforeseeable event.[2]
• "Errors as **slips and lapses** are the events which result from some failure in the execution and storage stage of an action sequence, regardless of whether or not the plan which guided them was adequate to achieve its objective."[2]
• Errors as **mistakes** are "deficiencies or failures in the judgmental and/or inferential processes involved in the selection of an objective or in the specification of the means to achieve it, irrespective of whether or not the actions directed by this decision-scheme run according to plan."[2]
• Something has been done that was not intended by the actor, not desired by a set of rules or an external observer, or that led the task or system outside its acceptable limits.[4]

Medicine
Medical definitions generally reflect Reason's[2] transdisciplinary definitions (see above) as we will see in more detail in the next chapter.

Definitions of Some Other Error-Associated Terms
Harm: Injury or damage to people, property, or environment.[14]
Danger: Injury, loss, pain, or other threat from an illness itself or contingencies not dependent on human (professional or patient) action.[15] N.B. This definition must be understood as a potential undesirable event which may, but not necessarily, happen in a particular situation; quoted for the sake of completeness only.
Risk: The result of error frequency combined with or multiplied by the severity of the consequences, resulting from its occurrence.[13] Events derived from doing something or not doing something (e.g., treatment, drugs, physical or mechanical action).[15]
Hazard: That error capable of causing harm; a potential source of harm.[14] (N.B. "Hazard" has a different meaning in clinical epidemiology). The most general term standing for both danger (from illness) and risk (from interventions).[15]

Continued

Table 2.1   Some Definitions of Error and Related Terms across Human Experience (*Continued*)

Error frequency: The resultant of exposure duration, the opportunities for error during that given exposure and the probabilities of detection and avoidance of harm during that exposure.[14]

Mistake: An error caused by a fault: the fault being a misjudgment, carelessness, or forgetfulness. [7] A repeated error is called a mistake.

Lexical: An error in action, calculation, opinion, or judgment caused by, for example, poor reasoning, carelessness, or insufficient knowledge.[16] Something the actor intended and that someone else did not intend; a mistake is an incorrect decision or choice or its immediate result, or an error in deciding what is intended; an error whose result was unintended, as opposed to an action that was intended; an incorrect intention, choice of criterion, or value judgment.[4] Error in planning, hence mental failure.[17]

Gaffe: A verbal (grammatical or literary) mistake, usually made in a social environment.[7]

Slip: An action not in accord with the actor's intention, the result of a good plan but a poor execution.[4] Syn.: An unintentional error. Potentially observable externalized actions-not-as-planned (slips of the tongue, slips of the pen, but also slips of action).[2] Error in execution, hence physical, sensory/motor failure.[17] Slips are associated with attention and perceptual failures and result in observable inappropriate actions whereas lapses are more cognitive events involving often memory failures.[18]

Lapse: More covert (than slips) error forms, largely involving failures of memory, that do not necessarily manifest themselves in actual behavior and may be apparent only to the person who experiences them.[2]

Both slips and lapses: Errors resulting from some failure in the execution or storage stage of an action sequence, regardless of whether the plan that guided them was adequate to achieve its objective.[2]

Violations: Deviations for safe operating practices, procedures, standards, or rules. Most violations are deliberate actions, even if sometimes they can be erroneous.[2,6,18]

Cognitive illusions: Phenomena that lead to a perception, judgment, or memory that reliably deviates from reality.[19] Therefore, they are errors in thinking, judgment, and memory.

Fault: Error carrying a pejorative connotation of blame or responsibility beyond that implied by the previous term.[4]

Table 2.1 Some Definitions of Error and Related Terms across Human Experience (*Continued*)

Accident: An unwanted or unwonted exchange of energy (in the case of physical damage). Some accidents are the consequence of error; some are not. A specific, identifiable, unexpected, unusual, and unintended event that occurs in a particular time and place, without apparent or deliberate cause but with marked effects. It implies a generally negative probabilistic outcome that may have been avoided or prevented had circumstances leading up to the accidents been recognized and acted upon, prior to its occurrence.[20] An unplanned event that interrupts the completion of an activity and that may (or may not) include injury or property damage.[21]

Incident: Events, processes, practices, or outcomes that are noteworthy by virtue of the hazards they create for, or the harm they cause, subjects (like patients in the medical world). (N.B. Incident reporting systems are meant to capture all incidents that are worthy of reporting.[22] All accidents are incidents, but not all incidents are accidents. An incident does not lead necessarily to harm to humans or property [partly from Busse[23]].)

Adverse effect: Unfavorable, undesirable, or harmful results.[24]

Adverse reaction: An undesirable or unwanted consequence of a preventive, diagnostic, or therapeutic procedure.[25]

Side effect: An effect, other than intended one, produced by preventive, diagnostic, or therapeutic procedure or regimen. Not necessarily harmful.[25]

Side effect: Any effect that occurs besides the main, expected, and desired one.

Fallacy: Any error of reasoning and decision making that contravenes logic and critical thinking based on valid and relevant evidence.[26] (For other definitions and extensive review of fallacies and cognitive errors in medicine, see elsewhere.)[22,26]

Human error: The failure of planned actions to achieve their desired ends without the intervention of some unforeseeable event.[2] An error committed by an individual.

Team error: Human error made in group processes. Mistakes and lapses are more likely to be associated with group processes.[27] A human error committed by individuals in a group within their interaction while planning, performing, and evaluating a task. A deficiency in knowledge and experience is the most frequent shared error.[27]

Violation: Deviations from safe operating practices, procedures, standards, or rules.

distinction between two views of relationship between errors and accidents, the Bad Apple Theory (an old view), or a new view, as summarized in Table 2.2.

Note regarding Error versus Adverse Effect

Adverse effects—meaning unfavorable, undesirable effects, some of them harmful—are not necessarily all based on error. Many of them cannot be originally anticipated. It is up to their observer to identify them and correct them. In an even broader context, as "side effects" they may include any effect that occurs besides the main expected and desired one. Side effects then include adverse effects.

Taxonomy of Error

Taxonomy is the practice and science of classification.[29] In our context, it is used to organize the diverse world of error. Human error may be classified simply as the assembly of different types of error without any order. Other classifications

Table 2.2   Two Views of Human Error according to Dekker[13]

The Old View	The New View
• Human error is a cause of trouble. • Human error is a cause of accidents as the dominant contributor to more than two-thirds of them. • Failures occur unexpectedly. • To explain failures, you must seek them.	• Human error is not a cause of trouble; it is a symptom or effect of some deeper system trouble. • Human error is not random. It is connected to other elements of the system (tools, tasks, environment). • Human error is not the conclusion but rather the starting point of an investigation. • To explain failure, do not try to find where people went wrong. • Identify people's inaccurate assessments, decisions, judgments. • Instead, find how people's assessments and actions make sense at the time, given the circumstances that surrounded them.

give some qualitative or quantitative direction to categories of events, so desirable in our practical dealings with human error. Categories are defined as clearly as possible with explicit inclusion and exclusion criteria; such criteria prevent misclassification of events and other units of observation. Linnaean taxonomy in biology, Bloom's taxonomy of learning, and educational objectives by levels of cognitive levels of complexity[30-32] or staging of cancer are directional; classifications of cancer by its type or site are mostly not.

In any case, any taxonomy should be usable, comprehensive, reliable, relevant, insightful, contextual, unambiguous, adaptable, useful, and authoritative.[33] That's where challenges of a good taxonomy lie and why these challenges are making some universal taxonomy of error next to impossible.

Taxonomies of human error in general may be directional or nondirectional, going from only two categories to many.

Person versus System

Errors may be classified or related simply as follows:

- **Person-oriented or human errors, personal factors**: an individual is at the origin of an error given his or her faulty knowledge, attitudes, or skills; deficient critical thinking and argumentation; and ultimately decision making. In addition to knowledge, reasoning, and decision-making skills, individual disposition, physiological and pathological states, physical or mental, or demographic and other characteristics of the reasoner, decision maker, and doer are important personal factors in relation to error.
- **System (functioning)-oriented errors or contextual factors** (human and material environment and interaction of their components are at the origin of error).[34] Errors affect either the structure and organization of an activity or its process (how things are done) and its impact (what it leads to in terms of various outcomes).
- **Due to some interaction between the two** as relationships between human factor and organizational failure.[35]

Most errors are of a human (person-oriented) nature[36]; others think the opposite, as discussed in Chapter 1. So far, there is no study in medicine that would decide which of those two informative and judgmental statements are true.

Planning versus Execution

Another classification recognizes two types of errors:[24]

- **Errors of planning**, stemming from an incorrect original intended action.
- **Errors of execution,** when the correct action does not proceed as intended.

In the error domain, any taxonomy must reflect and be usable for descriptive (occurrence), causal research, and intervention evaluation purposes.

Expertise, Its Quality, and Uses

Errors committed by human operators like workers, physicians, pilots, engineers, nurses, or others may be, according to Reason,[2] due either to the failure of or to the lack of expertise.

Cognition and Cognitive Process as a Core Source of Error and of Its Understanding and Control

Any accumulation of possible error generating human- and environment-related factors, stages or circumstances ultimately enter our judgment, critical thinking, and decision processes. Errors do not happen without ultimate failures of cognition and cognitive process. For example, it is easier to say that bad weather led to poor driving and a traffic accident than to specify why and how bad weather was cognitively processed and eventually misinterpreted and misused in error interpretation and management—the driver's decisions in this case.

But what is cognition and cognitive science? **Cognition** as a direct property of the brain means the process of awareness of thought or process of knowing.[37,38] The term refers to the faculty of processing information, applying knowledge, and changing preferences. In a more specific sense, it refers to the mental functions and processes (thought, comprehension, inference, decision making, planning and learning, capacities of abstraction, generalization, concretization/specialization, and meta-reasoning, beliefs, desiring, knowledge, preferences and intentions of intelligent individuals/objects/agents/systems). It is an information-processing outlook of mental functions in view of the development of knowledge and concepts within a group, culminating in both thought and action.[37,38]

Cognitive science as a multidisciplinary study of mind and behavior[39,40] is the product of contributions from psychology, psychiatry, philosophy, neuroscience, linguistics, anthropology, computer science, sociology, and biology.[41] It is used in the study of any kind of mental operation or structure what can be studied in precise terms.[42] In medicine, for example, the development, use, and evaluation of algorithms or decision-making processes or critical thinking analysis and evaluation in medical understanding and decision making all belong to the domain of cognitive science. In more general cognitive science, cognitive maps,[43] argument maps,[44,45] or concept maps[46] are visualizations of cognitive processes, their components, and interrelationships.

Cognitive processes are those processes involved in obtaining, storing, and retrieval of knowledge.[47] Communication is considered a primary cognitive function. The cognitive processes are remembering (recognizing and recalling), understanding (interpreting, exemplifying, classifying, summarizing, inferring, comparing, and explaining), applying (executing and implementing), analyzing (differentiating, organizing, and attributing), evaluating (checking and critiquing), and creating (generating, planning, and producing) functions.[48,49] Errors may be the result of any faulty function mentioned herein. In this light, even *reasoning* is seen by some as a sort of cognitive process in the search for reasons for or against beliefs, conclusions, actions, or feelings.[50,51]

Errors in cognitive processes are also called **cognitive biases** and are defined as pervasive tendencies for the information processing system to consistently favor stimulus material of a particular content over stimulus material of another content.[51] This definition overlaps to a variable degree with the notion of fallacies and error as well. We have listed and discussed cognitive biases in relation to fallacies in medicine elsewhere.[26,22,52]

In cognitive therapy, the term **cognitive distortion** is applied to at least 10 different aspects of cognition and reasoning, such as all-or-nothing thinking, overgeneralization, mental filtering, disqualifying the positive, jumping to conclusions, magnifying (catastrophizing, exaggerating) or minimizing (understating the magnitude of the reality), using emotional reasoning (emotions replacing reality), making *should* statements (what should be replaces reality) or having rigid (inflexible to circumstances) rules, labeling, and personalizing (attribution; replacing oneself for real reasons and reality).[53–55] These distortions are close in meaning to **heuristic biases** discussed more extensively later in this chapter.

Whether the source of error is *latent* (i.e., product of a system, mostly machine, function, interaction/organization of technical, environmental or social factors) or *active* (i.e., related to the working and decision-making individual),[18] cognition with its critical thinking and decision-making component is unavoidably an endpoint of whatever precedes it in the chain or sequence of reasoning, decisions, and right (error-free) or wrong (error-laden) actions taken. Decisions and their consequences are final filtered moments and steps of factors and functions sequences, processes, and systems that lead to them, their errors, and the consequences of the latter.

Models of Error, Their Development, and Contributing Sites and Entities in Context

Accident (meaning an event to which errors lead) models may be seen as the following:

- *Linear models* (i.e., sequences of events).
- *Systemic models,* similar to spatiotemporal and multidimensional webs of causes as they are discussed in epidemiology. Their evaluation focuses not only on each possible component of the web but also on the interactions between them.

Most models of environments (people, activities, equipment, and setting) are based on the *man–machine paradigm.* Such a paradigm includes plant, machine, tools, computers, and computer-based instruments with their monitors, keyboards, tracers, or printers in direct contact with the operator and the sociotechnical activity context in which the interaction of all these evolve.

Person-Oriented Models

Person-oriented models help us better understand where and why an error may occur on the path of cognition leading to the decision and action. Person-oriented errors are due either to failure of expertise or to lack of expertise, knowledge, reasoning, and decision making in general and in a specific setting. Beyond that, errors are related to a broader array of human factors

Errors are related to various *human factors* (ergonomics is a term sometimes used synonymously) that are defined most broadly as physical and cognitive properties of an individual or social behavior specific to humans and influencing the functioning of technological systems as well as of human–environment equilibriums. Their study aims at improving industrial designs, operational performance, and safety.[56] Cacciabue[18] defines human factors as those related to technology with the analysis and optimization of the relationship between people and their activities, by the integration of human sciences and engineering in systematic applications, in consideration for cognitive aspects and sociotechnical working contexts.

The term *human reliability* is related to the field of human factors in engineering, manufacturing, transportation, the military, or medicine. It depends among other human factors on age, circadian rhythms, state of mind, physical health, attitude, emotions, propensity for certain common mistakes, errors, and cognitive biases.[57] Properties of both reasoning and decision-making as well as actions reflect a person's characteristics. Human factors may be then viewed not only as factors related to the person but also as humans' interaction with their working and living environment.

Rasmussen's Model of Human Activity in Relation to Error

Person-oriented errors may occur at any stage or step on the path from activation, observation, identification, and interpretation to evaluation, goal selection, procedure selection, and action. For example, the directionality of these Rasmussen's steps[2,58,59] is also well reflected in our way of working up a diagnosis or making therapeutic decisions. In treating patients, we also proceed from observing them (at the bedside or at the laboratory) to making the diagnosis (interpretation) and proposing the treatment (deciding on a task to resolve the problem). More about this will follow in the next chapter.

Figure 2.1 illustrates Rasmussen's step-by-step path of adapted medical care from some kind of observation to its interpretation and ensuing task (action to solve the problem).[58] We may apply it here to medical care. Any of Rasmussen's steps in human functioning may be the subject of error. In a fairly similar direction and spirit, errors may occur at any step of Bloom's taxonomy of learning and teaching processes: [30-32]

- **Knowledge** (observation and recall of information).
- **Comprehension** (understanding information, translation into the specific context, interpretation, prediction of consequences).
- **Application** (uses of information or evidence, solving the problem).
- **Analysis** (identification of patterns, meaning of the whole and its components).
- **Synthesis** (generalizations, new ideas generation, merging knowledge from several areas, making conclusions and predictions).
- **Evaluation** (comparisons of and making discriminations between ideas, assessing the value of claims, making choices based on reasoned argument, verifying the value of evidence, making distinctions between subjectivity and objectivity).

Chapter 3 demonstrates that such directional structure of learning and general human activity may be applied to clinical activities and research as well, such as accepting the patient for care, evaluating an individual's risks of health problems, making a diagnosis, choosing treatment, establishing a prognosis, or writing orders and summaries in patient charts and files. Again, human and system errors may occur at any step and stage of clinical practice and medical research, and they really do. We are just one part of the general critically thinking community with all its strengths, weaknesses, and risks of errors.

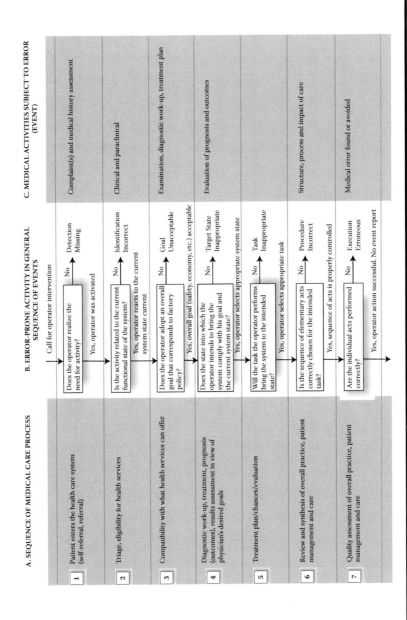

Figure 2.1 Activity-related errors. From manufacturing to medical activities and clinical care process. Based on Rasmussen's model of human functioning in general.[55,58]

Person-Related Errors in the Domain of Skills, Rules, and Knowledge

Given the nature of human activity, the *generic error-modeling system* (GEMS)[2] makes a distinction among the following:

- *Skill-based slips and lapses* as sensory-motor performances during acts or activities that, after a statement of an intention, take place without conscious control as smooth, automated, and highly integrated patterns of behavior.[59]
- *Rule-based mistakes,* in which wrong rules and their applications are the culprits.
- *Knowledge-based mistakes,* in which faulty reasoning and decision making based on poor evidence, ignoring alternatives, and other fallacies lead to error, even if the skills are correct and executed based on proper rules.

In the cognitive domain, errors established by psychologists include the following:[60]

- *False sensations* (discrepancies between the physician's world and the sensory apparatus).
- *Failures of attention.*
- *Memory lapses.*
- *Unintended words and actions.*
- *Recognition failures.*
- *Inaccurate and blocked recall.*
- *Errors of judgment* (psychological and temporal misjudgment, misconception of chance and covariation, misjudgment of risk, misdiagnoses, fallacies in probability judgments, and erroneous social assessments).

Physicians may commit any of these.

Models of Reasoning and Decision Making Related to Informal Logic and Critical Thinking: Aristotle, Toulmin, Heuristics

Errors in judgment, the last point in the previous list of errors as seen by psychologists, fall into two main categories of human problem and error solving:

1. The fast evolving, structurally, methodologically and in their applications expanding domain of "complete" modern informal logic, critical thinking, and decision making. Reasoning and decision making in optimal conditions.

2. Their parallel, truncated way of use and practice of heuristics dictated by the reality of modern-day pressures, ensuing limitations and alternatives as an additional option to "full-blown" thinking.

Argument and Argumentation Models in Optimal Conditions

Committing errors may be due not only to the use of poor evidence but also to the faulty use of evidence in argumentation and critical thinking as we see today in research and development and in practice. This approach to the error problem in general is still relatively novel and worthy of attention and practice.

Errors may occur and be explained by our ways of argumentation and the type, quality, and completeness of evidence used as a substrate and carrier of our premises and conclusions. This may be done either based on the classical Aristotelian categorical syllogism way of thinking or using Toulmin's modern way of argumentation,[61] relying in its conclusions (claims) on grounds, backing, warrants, qualifiers, and rebuttals as well as on the strength of evidence that underlies all of them. We have outlined and discussed those ways in more detail elsewhere.[62,63] (Other taxonomies of errors are also shown elsewhere.[4,Ch. 7])

Let us remind ourselves that ***categorical syllogisms*** operate on the basis of two premises: the first being specific to the particular case and the second reflecting a general situation related to the nature of the case. The conclusion of these premises shows the particular case in light of the general view of the problem under study: this driver, who became a traffic accident victim due to the error of passing, was drinking (premise A); drivers who drink cause traffic accidents related to errors in the assessment of a traffic situation (premise B); hence, the accident-related injury in the driver who erred by passing other vehicles was caused by drinking (conclusion).

Toulmin's model of argument is based on six elements. This model applies very well to various domains of human thinking and decision making including medicine and other health sciences.[62,64] Figure 2.2 illustrates Toulmin's way of reasoning.

The path from observations or data leads to some kind of claim. The *claim* stemming from the argument path is analyzed by exploring how information, such as observations or data related directly to the problem (*grounds*) may be seen in light of the additional past and present accumulated experience (*backing*) and of how our experience with data or other specific facts may be interpreted in view of the past experience (*warrant*), all three being linked together to offer some certainty about the claim (*qualifier*), provided that some exclusion criteria do not apply (*rebuttals*). The argumenter (rather than arguer) may err in working with poor-quality or otherwise inadequate grounds, backing, or warrant, and other errors may be committed by omitting or poorly considering relevant

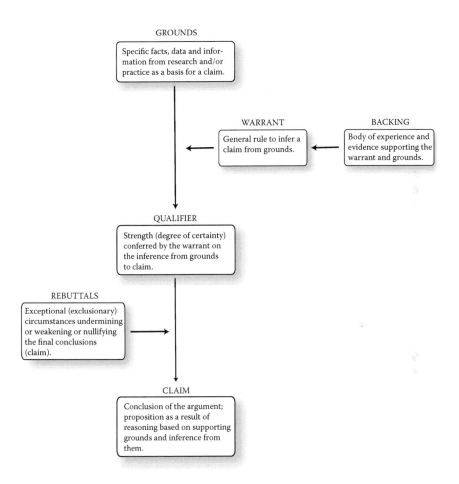

GROUNDS

Specific facts, data and information from research and/or practice as a basis for a claim.

WARRANT

General rule to infer a claim from grounds.

BACKING

Body of experience and evidence supporting the warrant and grounds.

QUALIFIER

Strength (degree of certainty) conferred by the warrant on the inference from grounds to claim.

REBUTTALS

Exceptional (exclusionary) circumstances undermining or weakening or nullifying the final conclusions (claim).

CLAIM

Conclusion of the argument; proposition as a result of reasoning based on supporting grounds and inference from them.

Figure 2.2   Components and structure of the modern argument as developed by S.E. Toulmin. Redrawn with modifications from Toulmin, S.E.; Jenicek, M.[61–64]

rebuttals. In addition, making links between argument building blocks may be erroneous themselves. Human error may then also be seen as a product of faulty argumentation. From the point of view of error, sites of erroneous decisions and ensuing actions may be considered happenings at the level of claims (conclusions, decisions, or claims) of an argumentation process. Premises may also be based on poor, wrong, or inexistent evidence, and the links between argument building blocks (premises) and the final claim may be erroneous or questionable. In our example, driving too fast (claim) may be based on poor grounds (data about driving conditions), lack of information about the past experience with such conditions (backing), unclear understanding, and explanation of the role of such experience in relation to driving conditions (warrant), giving a probability of the consequence of driving fast (qualifier) provided that some preventive measures have not been undertaken (rebuttals).

Until now, in the general error domain, possible erroneous logic and critical thinking behind human decisions and ensuing error and its consequences have been raised across the literature as possible important sources of error. However, we have not yet found in the literature available to us the applications and uses of modern critical thinking and informal logic to error identification, localization, explanation, correction, or prevention in particular error situations.

Heuristic Reasoning and Decision Making in Constraints of Time and in Urgency

In many situations (e.g., aviation, other modes of transportation, industrial productions, chess matches, sports, horse racing, the casino), there is often not enough time for more complete reasoning, full-blown argumentation, and decision-making processes. This reality, so well known to health professionals in surgery, obstetrics, or emergency medicine, requires adaptations that are a fertile ground for error. The domain of heuristics was established to better understand this kind of challenge.

Heuristics may be seen as "quick and dirty mental shortcuts to discovery, which sometimes err."[62] Heuristics is then seen as a method of rapidly coming to a solution that is close to the best possible answer, or optimal solution. Heuristics are "rules of thumb,"[65] educated guesses based on experience, intuitive judgments, or simply common sense.[66] Some kind of intelligent guesswork replaces more structured and complete ways of reasoning and decision making. The *Journal of Heuristics* is devoted to the methodology of heuristics and its application to business, engineering, and societal problem solving.

In computer science, heuristics are used to derive the most appropriate solution of several options found by alternative methods, which is then selected at successive stages of a program for use in the next step.[67] This process, different from the aforementioned definition, reduces the complexity of computational tasks. "Speeding things up" is a common trait of both concepts.

In opposition to heuristics, path analysis, algorithm developments, and their uses (e.g., decision-tree development and uses) avoid the trial-and-error approach as much as possible and offer a more justified and structured process to the problem solver.[68] It should be noted that some algorithms may be based on heuristics. In medicine, making a diagnosis at the bedside or in family practice often involves heuristic processes in the way they reduce the number of options in differential diagnosis.

Peter and Noreen Facione in their original coverage of links among thinking, reasoning, argumentation, and heuristics[45] offer for our consideration a list of 14 heuristic processes together with the reasoning disadvantages of each of them. Such imperfect heuristic processes may generate errors in medical practice and research as well. Table 2.3 summarizes heuristic maneuvers and their reasoning disadvantages.

Hindsight and rules of thumb, which are very often used in transportation and also in diagnostic and therapeutic decisions in family and emergency medicine, also belong to the heuristics domain. Some erroneous conclusions stem from *hindsight*, a way of reasoning and drawing conclusions about an event after it has happened.

Recognition of the realities, possibilities, or requirements of a situation may be influenced to a variable degree by what we have already seen, experienced, and understood in another and seemingly related or similar context. The *availability heuristic* hindsight-based cognitive error occurs if an observer bases his or her prediction of the frequency and understanding of other characteristics of an event on how easily an example can be brought to mind.[69] The "If you can think of it, it must be important" or "I knew it all along" effects are behind such *hindsight bias* or *vaticinium ex eventu* committed by a reasoner who overestimates the predictability or obviousness of an answer compared with the estimates of subjects who must guess without advance knowledge.[70] The vividness and emotional impact of recollection takes over the real probability of the new event under consideration. For example, hindsight bias matters in legal cases where a judge or jury must determine whether a defendant was legally negligent in failing to foresee a hazard. Hindsight bias systematically distorts the reality, giving more probability to the event than it actually has. Machine or vehicle operators and their errors and other outcomes may be seen and evaluated with some degree of hindsight bias.[70] They should not be. Examining possible alternatives may reduce the effects of this bias.[71]

Another inaccuracy in heuristic reasoning and decision making is related to the rule of thumb, in which estimations are made according to some rough and ready practical rules instead of on the basis of exact measurement and the rest of the scientific methods and its findings.[72] Sir William Hope, in the 17th century, stated, "What he doth, he doth by the rule of thumb, and not by art." This relates

Table 2.3 Heuristic Maneuvers and Their Reasoning Disadvantages

Name	Cognitive Maneuver	Disadvantage/Risk
Satisificing and temporizing	Given an option that is good enough, decide in favor of that option.	Good enough may not be best.
Affect	Take an initial stance in support of or in opposition to a given choice consistent with one's initial affective response to that choice.	Feelings may mislead.
Simulation	Estimate the likelihood of a given outcome based on one's ease in imagining that outcome.	Overestimation of one's chance of success or likelihood of failure.
Availability	Base the estimate of the likelihood of a future event on the vividness or ease of recalling a similar past event.	Mistaken estimations of the chances of events turning out in the future as they are remembered to have turned out in the past.
Representativeness—analogical	Infer that because this is like that in some way or other, this is like that in relevant ways.	The analogy may not hold.
Representativeness—associational	Connect ideas on the basis of word association and the memories, meanings, or impressions they might trigger.	Jumping from one idea to the next absent from any genuine logical connection and drawing inaccurate inferences from the combined thought process.
Generalizing from one to all	From a single salient instance draw a generalization about an entire group.	The one may not be representative of many.

Table 2.3 Heuristic Maneuvers and Their Reasoning Disadvantages (*Continued*)

Name	Cognitive Maneuver	Disadvantage/Risk
"Us vs. them" dynamic	Reduce problems to a simple choice between two opposing forces.	Conflict that excludes reasonable compromise.
"Master–slave" power differential	Accept without question a problem as presented by or a solution as proposed by a superior authority.	Working on the wrong problems, applying a mistaken solution.
Anchoring with adjustment	Having made an evaluation, adjust it as little as needed in light of new evidence.	Failure to reconsider thoroughly.
Control (Illusion of)	Estimate the level of control you have over the actual outcome of events, the amount of desire or the energy you put into trying to shape those events.	Overestimation of one's power to control events or underestimation of one's actual responsibility for what happened.
Elimination by aspect	Eliminate an option or group of options from consideration upon the discovery of an undesirable failure.	Failure to give full holistic consideration to viable options.
Risk and loss aversion	Avoid the foreseeable risk of sustaining a loss by not changing the status quo.	Paralysis of decision making stuck in the deteriorating status quo.
Zero-out tendency	Simplify decision contexts by treating the remote probabilities as if they are not even possibilities.	Failure to appreciate the possibilities that events could actually turn out differently than expected.

Source: From Facione, P. A. and Facione, N. C., *Thinking and Reasoning in Human Decision Making: The Method of Argument and Heuristic Analysis,* The California Academic Press LLC, Milbrae, CA, 2007. With permission.

to approximations of ancient laws like the one authorizing a man to legally beat his wife, provided that he use a stick not thicker than his thumb.[73] The reference to thickness of the thumb is not specified. Testing the temperature of cooking by immersing the cook's thumb in the liquid, parents testing the temperature of milk before feeding their children, a tailor's judgment that "twice around the thumb is once around the wrist," or considering the distance from the tip of the nose to the outstretched fingers as about 1 yard[74] are all some kind of rule of thumb.

Physicians in any stressful and precipitous situations requiring vital instant decisions must make decisions based on rules of thumb, hindsight (biased or not), and other heuristically laden ways of decision making. This rule is now under scrutiny both in emergency and family medicine.[75] Without analyzing and understanding individual behavior, it is hard, if not impossible, to understand the nature of human error.

All those models of possibly erroneous reasoning occur in the broader context of additional environments and their components. They may cumulate into situations that may lead to human error. The error of an individual happens in relation to the environment in which it occurs. The system-oriented approach to error understanding and control reflects this reality.

System Functioning-Oriented Models, or "One Thing Goes with and Leads to Another"

Dekker[28] recognizes three models of "accidents" (or errors in our sense):

- The *sequence-of-events model* views accidents as a chain of events leading to errors and harm.
- The *systemic model* sees errors and ensuing accidents as a product of the interactions between system components and processes rather than failures within them. The systems works normally, but people and organizations use imperfect knowledge under various kinds of pressures of functioning like time, competition, and alike.
- The *epidemiological model* pays attention to latent failures as "pathogens" that do not normally wreak havoc unless they are activated by other factors. We will see examine the epidemiological view of error in medicine in a more traditional way in Chapter 3, in the context of a web of causes and web of consequences. Some immediate comments follow in the next main section of this chapter.

Hence, one of the alternative views of the error problem is by looking at it as a *system problem*. A system is a set of interacting and interdependent entities, real or abstract, forming an integrated whole. In management science, operational research, or organizational development, human and physical organizations are viewed as systems of interacting components and their aggregates, carriers of numerous processes, and organizational structures.[76] For example, algorithms are direction-giving systems, while decision trees are direction-searching ones. (More about this will follow in the next chapter). *Systems thinking*[77] tries to understand the system (and what the system represents) focusing on the analysis and interpretation of links and interactions between the system components. Obviously, a lot depends on how the system and its components are a priori defined. A *human–machine system* or *man–machine system* looks at the human operator, machine, and the related environment as an integrated entity whose understanding is essential in the given problem solving. Applied psychology, ergonomics, cognitive ergonomics, human computer–human machine interaction, or user experience engineering, physiology, computer sciences, and engineering all contribute to the understanding of man–machine system as a potential or real generator of error and its solution.[78]

In medicine, for example, we may consider a surgeon performing laparoscopic intervention, an orthopedic surgeon doing a hip replacement, or a cardiologist catheterizing for angioplasty together with their armamentarium as a man–machine system, or eventually, with their patient, as a man–machine–man system that may be analyzed for their successes and errors by human–machine methods and a broader experience is acquired also beyond the health science field.

Figure 2.3 illustrates that the generation of an error may occur at any stage of the hierarchy starting with some failure in the individual or collective environment cascading or interacting in its causes (or not) in a broader or strictly individual or collective context. The way we see human activity, its context, and related factors will largely determine our understanding of it, our research about it, and our definition of measures and actions as remedies to human error making.

In summary, human thinking and activity are not the only sources of error. Such individual and collective processes occur in some general physical environment, working milieu, or social domain (family, learning, or leisure setting), cultural, historical, science, and faith domains, technologies uses situations and material context as well. Human error can be considered a failure in any of these contexts as well as occurring in the chain and path of interactions between all of the aforementioned domains.

A Practical Example of an Erroneous Event and of Its Steps as Seen through Their Identification in Various Taxonomies of Error

In Figure 2.4, let us consider a sequence of events that lead Peter from the loss of his job to causing a traffic accident and the subsequent injury of an innocent

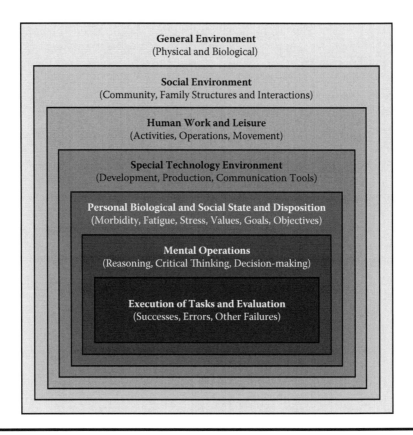

Figure 2.3 A cascade of error-related factors from general environment to the execution of tasks and its evaluation. Modified and redrawn from Zhang, J. et al., *J Am Med Inform Assoc*, 9, 6, 2002.

bystander. According to different paradigms of error, Peter's job loss may be considered a root cause of his accident and its consequences.

An Epidemiological Approach to the Error Problem

In health sciences, more than elsewhere perhaps, we are accustomed to seeing health phenomena as occurring in various frequencies from sporadic cases to epidemic, pandemic, or endemic events. Human error, especially in health sciences, must be seen also in this light if we wish to properly describe its occurrence in relation to its persons/time/place characteristics to explain its causes or to effectively control or prevent undesirable events or to enhance the desirable

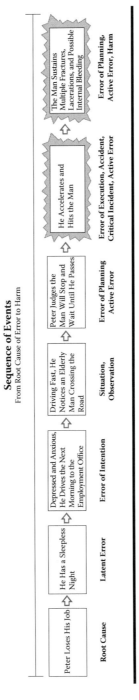

Figure 2.4 Sequence of events from the root cause to error happening and ensuing harm.

ones. Descriptive, observational analytical, and experimental/interventional epidemiology applies! We will discuss the "epidemiology of error" in the next chapter. In this one, let us retain as an enriching experience from the general error domain the root cause analysis (RCA), particularly interesting in situations in which epidemiological and biostatistical minds may feel disarmed in the study and interpretation of less frequent, if not exceptional cases, as error events may be.

A Word about Root Cause Analysis and Research

Errors, like disease in the health domain, may be viewed as having at their origin a web of causes in a particular sequence in time and interrelated distribution in space. In those webs, in the spirit of John Stuart Mill's philosophy, the necessary causes (i.e., the ones required absolutely for something else to occur) and sufficient causes (i.e., those sets of factors, conditions, or events that are needed to produce a given outcome regardless of what other conditions are present or absent) are considered. RCA tries to identify a kind of necessary cause, a kind of triggering element leading to a chain of events as an accumulation of error eventually causing an accident or any other undesirable event. This analysis is of particular interest for the study of rather sporadic serious problems with serious consequences (accidents). This analysis is of particular interest for the study of rather sporadic serious problems with serious consequences (accidents). The problem of sporadic cases analysis is magnified by the fact that sporadic cases involve some causal link between them (like common exposure to a given source of transmission between cases of infection or spread of mass psychological phenomena) that has not yet been established.

A root cause is an initiating cause of a causal chain that leads to an outcome of effect of interest—an element where an intervention could reasonably be implemented to change performance or prevent an undesirable outcome.[81] A root cause analysis is a class of problem-solving methods aimed at identifying the root causes of problems or events.[82] It is a procedure whose purpose is to ascertain and analyze the causes of problems, to determine how these problems can be solved or be prevented from occurring. Achieving the permanent resolution of the problem is its ultimate goal.[83]

In essence, it is a retrospective reconstruction of events leading to an error and its consequences. Root cause analysis depends heavily on some degree of a priori knowledge or supposition of possible or real causal factors. It is not a substitute for causal proof by stating that "this is a cause" instead of focusing on confirmation or rejection of classical criteria of causality as available across the philosophical and medical literature[68] in this or that particular case.

The aim and procedure of such an analysis is to do the following:

■ Reconstruct the whole event of error development in its time–space sequence.
■ Link elements in such developments to error and its consequences in a plausible causal and logical network.
■ Identify in the best possible way the role of each incident in the causal chain and propose in a hierarchical way events that should be corrected to control the ensuing error and its consequences.

It is a kind of "going for the jugular" in the error's web of causes.

The "five whys" in the proof consist of the following:

■ that some events as presumed cause exist.
■ that such events lead to a stated event (error and its consequences).
■ of demonstration that such a link between events exists.
■ and check that possible other contributing events (causes) are needed for the problem to occur.
■ that the causal demonstration considers alternative explanations—that is, looking to see if anything else might contribute to the event and might eventually be a better explanation and the risk of such alternatives.[84]

Root cause analysis based on some kind of cause–consequence tree assessment attempts to reconstruct relevant sociotechnical factors possibly at the core of inappropriate behaviors, system failures, and malfunctions. It also makes use of models of cognition and taxonomies. The reconstruction is an open-ended process whose end (what is a primitive important "cause") is left up to the analyst.

In the world of medicine, root cause analysis and its results are not unanimously accepted and appreciated. More about this will follow in Chapter 5.

Beyond Epidemiology: Other Models of Search for Causes

For Cacciabue,[18] three models and methods might be of interest:

1. Analyzing four cognitive functions behind the event of interest: perception, interpretation, planning, and execution; a CREAM model.
2. Establishing a taxonomy of accident/incident data reporting (ADREP) in a timely fashion.
3. Some kind of integrated systemic approach for accident causation (ISAAC) in which both system factors and human factors pathways and their failures are considered.

Epidemiological Implications of the Error Analysis Problem

Persisting diversity of fundamental definitions and of taxonomy makes the study of error challenging, including data collection, epidemiological surveillance, or error causes. What kind of error, what kind of error causes, and what error settings do our studies focus on?

Classical criteria of cause–effect relationship cannot always be applied here given the frequent relative rarity of analyzed events, lack of operational definitions, and unclear taxonomy of both presumed causes and their consequences in the web of causes and web of consequences context. Root cause analysis also requires at least some a priori anticipated and presumptive cause that should be confirmed to some degree by the root cause analytical process.

Root cause analysis was proposed, used, and applied in health sciences, as we will see in the next chapter.[85,86] For example, a fluorouracil incident was the subject of root analysis in a recent Canadian experience.[87] However, root cause analysis is an analysis of a linearly chained events rather than an analysis based on a web of causes in their time and space relationships on which path analysis and other models of analysis are based.[70]

All in all, webs of causes and webs of consequences in the error domain require working with multidimensional models of error generation and consequences as well as respecting the diversity of domains to which they belong and apply. Those models are often specific to particular domains of human activity such as industry, transportation, information, or other new technologies, and they "do not travel well" in their uses and interpretation of findings from one domain to another.

Multivariable (several causes in focus) or multivariate (several consequences under study) methods of analysis are, for the moment, limited given challenges of taxonomy and statistical considerations, as already stressed. In this spirit, epidemiological methods so far remain rooted mainly in the health domain. More about this will follow in the next chapter.

Experimental epidemiology involves analytical studies by controlled means of exposure, constitution of control groups, and better a priori and a posteriori minimization of confounding and bias than in observational research. Randomized or otherwise controlled trials may be impossible due to the nature of the error problem and its setting, ethical considerations, or time, material, and human resources needed. Given such circumstances, the thought experiment concept is one possible alternative and compromise to the trial methodological rigor, taking into account the limited uses of experimental designs in the human error domain.

Thought Experiment: A Complement to Epidemiology?

A *thought experiment* (Ger. *Gedankenexperiment*, since 1820, or experiment conducted in the thoughts)[88] is a sort of "what if" reflection. It is a proposal for an experiment (by way of an intellectual exercise) that would test a hypothesis or theory but cannot actually be performed due to practical (e.g., means, ethics, feasibility) limitations. Instead, its purpose is to explore in a reflective way the potential consequences of the principle in question. In the error domain, experimental proofs are very limited given the most fundamental ethical considerations related to the error consequences manipulations. However, thought experiments may be attractive options in the future. So far, current experience and methodologies remain limited in the error domain.

In a thought experiment, data are formulated into a truth table where researchers ask themselves a question. If the answer is yes, another question is raised, and so on. Any no's means going back to a preceding yes step and proceeding in a modified manner.

Many philosophers prefer to consider thought experiments as merely the use of a hypothetical (i.e., imaginary) scenario to help understand the way things actually are.[88] It is some kind of trade-off between reality and its substitution to some situation that is still feasible to evaluate given the circumstances. Thought experiments display a patterned way of thinking that allows us to explain, predict, or control events in a more productive way than with some more haphazard alternative ways of handling the problem.

In science, thought experiments are "proxy" experiments.[88] However, wherever relevant, they must follow the same causal criteria as in any other observational, analytical, or experimental research. More about this can be found summarized elsewhere in the literature.[82]

Implications in the Search for Understanding, Control, and Prevention of Error Today

Given the aforementioned diversity in definitions, taxonomy, frequency, and nature of error, occurrence and etiological research differs from the established path of the scientific method or epidemiological and clinical research. Causes of error are most often based on speculative assessments of error generating situations; however, such speculations may be right or wrong from one case to another.

In the Research Domain

On the basis of existing error models and taxonomies and to answer error questions related to them, anyone in search of error understanding will be expected to answer following questions:

- What is the purpose of the research activity? To describe what happens, to explain causes of such happening, to assess what might be the most effective way to prevent or control a given error problem and its impact?
- How do we define and classify in formal and operational terms all variables of interest such as error, its causes, its outcomes, and characteristics of persons/time/place involved in the error event?
- Are we studying error as a simple event, as a result of some intervening causes, as being itself a cause of some (and which) outcomes?
- What distinctions are made between such entities as error, accident, harm, mistake, gaffe, slip, lapse, violation, fault, adverse effect, and side effect, and what is the reason for such distinctions in a given context of an error problem to be solved?
- What origin of error is studied: person-related, human/material system and interactions within such system, or some interaction between the two?
- Is the error of interest an error of planning or an error of execution?
- Is the error due to the failure of expertise or the lack of expertise?
- What are the moments and steps in the cognitive process and its teaching and learning equivalents whose faults are behind the error under study?
- What errors in the cognitive process, critical thinking, reasoning, logic, or argumentation have been made and have led to faulty claims, conclusions, or recommendations that led to error?
- Are good or bad heuristic processes contributing to the error development? If so, which ones are they?
- What models of error development may and should be established: linear sequences of events or webs of causes in their proper sequence and persons/time/place relationship?
- Which phenomena in such error developments may be considered latent causes, and which are the active (individual related) ones?
- What established epidemiological research methods of description, analysis, or intervention should be used in each particular error case?

There are no ever-encompassing single studies offering explanations of error in general. Instead, they offer better answers to some questions within such a general framework.

Figure 2.5 illustrates how the error problem might be fragmented into manageable pieces or partial questions leading to more tangible and concrete answers. Better answers and solutions to error problems may be obtained if the error problem is within a particular domain of human activity such as research and development, manufacturing, transportation, military arts, administration, and management or health sciences and alike. In each human activity domain, error may be caused by deficient knowledge, attitudes, skills, or critical thinking of active individuals. The deficiencies of these individuals may lead to unsound, poorly structured, erratic processes, all having a possible role in the erroneous impact of what was done. Even in the error domain, small is beautiful, as the popular saying goes.

In the Control and Prevention Domains

Given the already mentioned general methodological nature of the error problem, difficulty and limitations of the knowledge of a priori causes, and frequent uniqueness of events, any surveillance of error has its inherent limits. Ethical reasons limit experimental approaches, control groups, application of "contrasts" (alternative methods and processes), and inferentially acceptable results of such comparisons. Knowing incidence rates of errors is not enough; even obtaining them is not always possible.

We still do not know the strength and the specificity of prevention/control interventions and their expected control of error and how effective, efficacious, and efficient they are in terms of the evaluation methodology. Recent methodological developments and increasing frequency and experience in the domain of qualitative research, case studies and case research, or mixed methods research are encouraging us to use them according to their best content, orientation, and extent in the human error domain. Such experience also remains limited for the moment.

Conclusions: Ensuing State of the Human Error Domain Today

Experience in dealing with human error is rapidly expanding in both general and medical domains. However, more and more it is becoming evident that inherent methodological, factual, and ethical limitations require innovative approaches to the error problem. The practice of state-of-the-art scientific methodologies is often limited. Flexible and adaptable alternatives must complete the armamentarium of human error management.

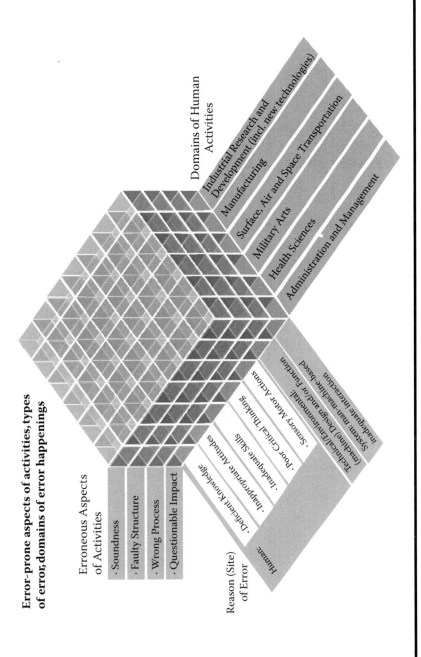

Figure 2.5 Error-prone aspects of activities, types of error, and domains of error happenings.

Given the limitations in the impact assessment, the error study focuses consequently with much more success on the structure or process analysis rather than on the error impact in its role of either cause or consequence (causes of error vs. error as cause of something else like accidents or health or social damage). Let us not forget that errors may then be studied both as dependent and independent variables in various causal models or may be studied just as a simple event.[89]

Does human error in general fall into what we might call evidence-based lathology? In other evidence-based domains and practice like in medicine and nursing, evidence in the error domain also means any data or information, whether solid or weak, obtained through experience, observational, or experimental work, relevant either to the understanding of the error problem or the decision making about it. The hierarchy of evidence in the error domain has not yet been firmly established; what is the "best evidence" in the error context? We are looking here at the best evidence of what is wrong rather than at what is best—the latter so far prevailing in focus in health sciences.

The dilemma between individual and system error remains. The study, analysis, and understanding of human error has a choice of two strategies: either the problem is fractioned into its elementary parts, which are studied and analyzed one by one; or a complex system within which error occurs is treated as a whole, taking into account the interactions of its composing elements. Trying to better understand complex systems in which errors occur, Dörner[90] points out insufficient goal elaboration, insufficient formulation of hypotheses about the structure of the system, insufficient ideas about the behavior of the system in time, insufficient coordination of different measures, a "ballistic" action responsible for the lack of detection of wrong hypotheses and inappropriate strategies, and also no self-reflection, meaning no repair of wrong hypotheses and inappropriate strategies behind system failures. Both human (individual) and system errors require balanced attention and integration of experience at both levels of error. Methodology to do that is now available to all.

Now, let us "migrate" the error problem and challenge in the best possible way into the domain of medicine and other health sciences. The next chapter provides a general overview in a spirit similar to that of this chapter.

References

1. *The Study of Human Error.* Edited by J. Jastrow. New York: D. Appleton-Century Company, 1936.
2. Reason J. *Human Error.* Cambridge, UK: Cambridge University Press, 1990.
3. *New Technology and Human Error.* Edited by J Rasmussen, K Duncan, J Leplat. Chichester, UK: John Wiley & Sons, 1987.

4. *Human Error: Cause, Prediction, and Reduction.* Analysis and Synthesis by JW Senders, NP Moray. Hillsdale, NJ: Lawrence Erlbaum Associates, Publishers, 1991.

5. Jastrow J. *The Procession of Ideas.* Pp. 1–36 in Ref 1.

6. Reason J. *A Framework for Classifying Errors.* Pp. 5–14 in Ref 2.

7. Wikipedia, the Free Encyclopedia. *Error.* Available at: http://en.wikipedia.org/wiki/Errror, retrieved September 26, 2008.

8. Reese WL. *Dictionary of Philosophy and Religion: Eastern and Western Thought.* Amherst, NY: Humanity Books (An Imprint of Prometheus Books), 1999.

9. Error. *The American Heritage® Dictionary of the English Language, Fourth Edition.* Houghton Mifflin Company, 2004. Available at: http://dictionary.reference.com/browse/error, retrieved July 21, 2009

10. Error. *The American Heritage® Stedman's Medical Dictionary.* Houghton Mifflin Company, 2002. Available at: http://dictionary.reference.com/browse/errror, retrieved July 21, 2009.

11. Clapp JE. *Random House Webster's Dictionary of the Law.* New York: Random House, Inc., 2000.

12. "*Error.*" Encyclopaedia Britannica. *Ultimate Reference Suite.* Electronic Edition. Chicago: Encyclopaedia Britannica, 2008.

13. Dekker SWA. *Errors in our understanding of human error: the real lessons from aviation and healthcare.* Technical Report 2003-01. Lund: University School of Aviation, 2003.

14. Peters GA, Peters BJ. *Human Error. Causes and Control.* Boca Raton, FL: CRC/Taylor & Francis Group, 2006.

15. Fagerhaugh SY, Strauss A, Suczek B, Wiener CL. *Hazards in Hospital Care. Ensuring Patient Safety.* San Francisco: Jossey-Bass Publishers, 1987.

16. *Random House Webster's Unabridged Dictionary 3.0.* Electronic Edition. Amsterdam: Random House, 1998.

17. Cosby KS. Developing taxonomies for adverse events in emergency medicine. Chapter 5, pp. 58–69 in: Croskerry P, Cosby KS, Schenkel SM, Wears RL. *Patient Safety in Emergency Medicine.* Philadelphia: Wolters Kluwer | Lippincott Williams & Wilkins, 2009.

18. Cacciabue PC. *Guide to Applying Human Factors Methods. Human Error and Accident Management in Safety Critical Systems.* London: Springer Verlag, 2004.

19. *Cognitive Illusions. A Handbook on Fallacies and Biases in Thinking, Judgment and Memory.* Edited by RF Pohl. Hove: Psychology Press (Taylor & Francis Group), 2004.

20. Wikipedia, the Free Encyclopedia. *Accident.* Available at: http://en.wikipedia/org/wiki/accident, retrieved November 11, 2008.

21. CCOHS. *Accident investigation. What is an accident and why should it be investigated?* Ottawa: Canadian Centre for Occupational Health and Safety, April 20, 2006, version. Available at: http://www.ccohs.ca/oshanswers/hsprograms/investig.html, retrieved November 11, 2008.

22. Croskerry P, Cosby KS, Schenkel SM, Wears RL. *Patient Safety in Emergency Medicine.* Philadelphia: Wolters Kluwer | Lippincott Williams & Wilkins, 2009.

23. Busse DK. *Cognitive Error Analysis in Accident and Incident Investigation in Safety-Critical Domains.* A PhD Thesis. Glasgow: University of Glasgow, Department of Computing Science, September 2002. Available at: http://www.dcs.gla.ac.uk/~johnosn/papers/Phd_DBusse.pdf, retrieved October 25, 2008.

24. *Encarta® Webster's Dictionary of the English Language.* Second Edition. A Bloomsbury Reference Book. New York: Bloomsbury Publishing Plc., 2004.

25. *A Dictionary of Epidemiology.* Fifth Edition. Edited by M Porta, S Greenlad, JM Last. A handbook sponsored by the I.E.A. Oxford: Oxford University Press, 2008.

26. Jenicek M. *Fallacy-Free Reasoning in Medicine. Improving Communication and Decision Making in Research and Practice.* Chicago: American Medical Association Press, 2009.

27. Sasou K, Reason J. Team errors: definition and taxonomy. *Reliability Eng Syst Safety,* 1999; **65**(1):1–9.

28. Dekker S. *The Field Guide to Understanding Human Error.* Aldershot: Ashgate Publishing Limited and Ashgate Publishing Company, 2006.

29. Wikipedia, the Free Encyclopedia. *Taxonomy.* Available at: http://en.wikipedia.org/wiki/Taxonomy, accessed October 23, 2008.

30. Bloom BS. *Taxonomy of Educational Objectives. Handbook I: The Cognitive Domain.* New York: David McKay Co Inc., 1956.

31. Anderson L, Krathwohl DR, Airasian PW, Cruikshank KA, Mayer RE, Pintrich PR, et al. (Eds.). *A Taxonomy for Learning, Teaching and Assessing: A Revision of Bloom's Taxonomy of Educational Objectives.* New York: Addison, Wesley, Longman Inc., 2001.

32. Forehand M. *Bloom's Taxonomy.* From *Emerging Perspectives on Learning, Teaching and Technology.* Available at: http://projects.coe.uga.edu/epltt/index.php?title=Bloom%27s_Taxonomy, retrieved October 23, 2008.

33. Cosby KS. Developing taxonomies for adverse events in emergency medicine. Pp. 58–84 in Ref.20.

34. Reason J. Human error: models and management. *BMJ,* 2000; **320** (March 18): 768–70.

35. Shappell DA. A human error approach to accident investigation: The taxonomy of unsafe operations. *Int J Aviation Psychol,* 1997;**7**(4):269–91.

36. Johnson C. *Visualizing the Relationship between Human Error and Organizational Failure.* Chapter 9 of the C Johnson's on-line *Handbook of Accident and Incident Reporting.* Available at: http://www.dcs.gla.ac.uk/~johnson/papers/fault_trees/organisational_error.html, retrieved November 9, 2008.

37. Wikipedia, the Free Encyclopedia. *Cognition.* Available at: http://en.wikipedia.org/wiki/Cognition, retrieved November 12, 2008.

38. Answers.com. *Cognition.* Available at: http://www.answers.com/topic/cognition, retrieved November 12, 2008.

39. Luger G. *Cognitive Science: The Science of Intelligent Systems.* San Diego, CA: Academic Press, 1994.

40. NationMaster.com. *Encyclopedia > Cognitive science.* Available at: http://www.nationmaster.com/encyclopedia/Cognitive_science, retrieved November 2, 2008.

41. Wikipedia, the Free Encyclopedia. *Cognitive science.* Available at: http://en.wikipedia.org/wiki/Cognitive_science, retrieved November 12, 2008.

42. Lakoff G, Johnson M. *Philosophy in the Flesh.* New York: Basic Books, 1999.

43. Wikipedia, the Free Encyclopedia. *Cognitive map.* Available at: http://en.wikipedia.org/wiki/Cognitive_map, retrieved May 12, 2008.
44. Wikipedia, the Free Encyclopedia. *Argument map.* Available at: http://en.wikipedia.org/wiki/Argument_map, retrieved May 12, 2008.
45. Facione PA, Facione NC. *Thinking and Reasoning in Human Decision Making: The Method of Argument and Heuristic Analysis.* Milbrae: California Academic Press LLC, 2007.
46. Wikipedia, the Free Encyclopedia. *Concept map.* Available at: http://en.wikipedia.org/wiki/Concept_Map, retrieved May 12, 2008.
47. WordNet-Online. *Basic Cognitive Process.* Available at: http://wordnet-online.com/basic_cognitive_process.shtml, retrieved November 2, 2008.
48. Fredson G. *5.1. The Cognitive Process Dimension.* Available at: http://www.udel.edu/educ/gottfredson/451/bloom.pdf, retrieved November 12, 2008.
49. Smith BC. *Cognition.* Pp. 138–139 in: *The Oxford Companion to Philosophy.* Edited by T Honderich. Oxford: Oxford University Press, 1995.
50. Wikipedia, the Free Encyclopedia. *Reasoning.* Available at: http://en.wikipedia.org/wiki/Reasoning, retrieved May 12, 2008.
51. Wikipedia, the Free Encyclopedia. *Cognitive bias.* Available at: http://en.wikipedia.org/wiki/Cognitive_bias, retrieved February 3, 2008.
52. Wikipedia, the Free Encyclopedia. *Category: Cognitive biases.* Available at: http://en.wikipedia.org/wiki/category_Cognitive_biases, retrieved January 27, 2008.
53. Wikipedia, the Free Encyclopedia. *Cognitive distortion.* Available at: http://en.wikipedia.org/wiki/Cognitive_distortion, retrieved January 28, 2008.
54. Beck AT. *Cognitive Therapy and the Emotional Disorders.* International Universities Press, Inc., 1975.
55. Burns DD. *The Feeling Good Handbook.* New York: William Morrow and Comp. Inc., 1989.
56. Wikipedia, the Free Encyclopedia. *Human factors.* Available at: http://en.wikipedia.org/wiki/Human_factors, retrieved November 17, 2008.
57. Wikipedia, the Free Encyclopedia. *Human reliability.* Available at: http://en.wikipedia.org/wiki/Homan_error, retrieved November 17, 2008.
58. Rasmussen J. Human errors: A taxonomy for describing human malfunction in industrial installations. *J Occup Accidents*, 1982;**4**:311–33.
59. Rasmussen J. *Information, Processing and Human-Machine Interaction.* Amsterdam: North-Holland, 1986.
60. Reason J. A framework for classifying errors. Pp. 5–14 in Ref 3.
61. Toulmin SE. *The Uses of Argument. Updated Edition.* Cambridge, UK: Cambridge University Press, 2003. (First edition, CUP, 1958.)
62. Jenicek M, Hitchcock DL. *Evidence-Based Practice. Logic and Critical Thinking in Medicine.* Chicago: American Medical Association Press, 2005.
63. Jenicek M. *A Physician's Self-Paced Guide to Critical Thinking.* Chicago: American Medical Association Press, 2006.
64. Toulmin S, Rieke R, Janik A. *An Introduction to Reasoning.* 2nd Edition. New York: Collier Macmillan Publishers, 1984.
65. Wikipedia, the Free Encyclopedia. *Heuristic.* Available at: http://en.wikipedia.org/wiki/Heuristic, retrieved November 5, 2008.

66. Answers.com. *Heuristic.* Available at: http://www.answers.com/topic/heuristic, retrieved November 5, 2008.
67. Wiktionary, a Free Dictionary. *Rule of Thumb.* Available at: http://en.wiktionary. org/wiki/rule_of_thumb, retrieved November 5, 2008.
68. Jenicek M. *Foundations of Evidence-Based Medicine.* Boca Raton, FL: The Parthenon Publishing Group Inc. (CRC Press), 2003.
69. Wikipedia, the Free Encyclopedia. *Availability heuristic.* Available at: http:// en.wikipedia.org/wiki/Availability_heuristic, retrieved December 18, 2008.
70. The BizOp News. *Due Diligence: What Is Hindsight Bias?* Available at: http://bizop. ca/blog2/due-diligence/hindsight-bias.html, retrieved December 19, 2008.
71. Wikipedia, the Free Encyclopedia. *Hindsight bias.* Available at: http://en.wikipedia. org/wiki/hindsight_bias, retrieved December 18, 2008.
72. Phrases.org. *Rule of thumb.* Available at: http://www.phrases.org.uk/meanings/rule-of-thumb.html. retrieved December 18, 2008.
73. Wikipedia, the Free Encyclopedia. *Rule of thumb.* Available at http://en.wikipedia. org/wiki/Rule_of_thumb, retrieved December 18, 2008.
74. Quinion M. *World Wide Words. Rule of Thumb.* Available at: http://www.world-widewords.org/qa/qa-rul1.htm, retrieved December 18, 2008.
75. André M, Borquist L, Mölstad S. Use of rules of thumb in the consultation in general practice—an act of balance between the individual and the general perspective. *Fam Pract,* 2003;**20**(5):514–9.
76. Wikipedia, the Free Encyclopedia. *System.* Available at: http://en.wikipedia.org/ wiki/System, retrieved November 17, 2008.
77. Wikipedia, the Free Encyclopedia. *Systems thinking.* Available at: http://en.wikipedia. org/wiki/Systems_thinking, retrieved November 17, 2008.
78. Wikipedia, the Free Encyclopedia. *Human factors.* Available at: http://en.wikipedia. org/wiki/Human_factors, retrieved November 17, 2008.
79. Zhang J, Patel VL, Johnson TR. Medical error: Is the solution medical or cognitive? *J Am Med Inform Assoc,* 2002(Nov–Dec);**9**(6):S75–S77.
80. Zhang J, Patel VL, Johnson TR, Shortliffe EH. Toward an action based taxonomy of human errors in medicine. *Proceedings of the 24th Conference of Cognitive Science Society, 2002.* Pp. 934–8 (2003). Available at http://acad88.sahs.uth. tmc.edu/research/publications/cogsci2002-taxonomy.pdf, retrieved December 17, 2007.
81. Wikipedia, the Free Encyclopedia. *Root Cause.* Available at: http://en.wikipedia. org/wiki/Root_cause, retrieved November 10, 2008.
82. Wikipedia, the Free Encyclopedia. *Root Cause Analysis.* Available at: http:// en.wikipedia.org/wiki/Root_cause_analysis, retrieved November 10, 2008.
83. 12manage, the Executive Fast Track. *Root Cause Analysis.* Available at: http:// www.12manage.com/methods_root_cause_analysis.html, retrieved November 10, 2008.
84. Bill-Wilson.net. *Root Cause Analysis: RCA Particles. Five-by-Five Whys.* Available at: http://www.bill-wilson.net/b73.html, retrieved November 10, 2008.
85. Canadian Patient Safety Institute. *List of guides to patient safety.* Available at: http:// www.patientsafetyinstitute.ca/resources/tools.html, retrieved November 10, 2008.

86. Canadian Patient Safety Institute, Institute for Safe Medication Practices Canada, Saskatchewan Health. *Canadian Root Cause Analysis Framework. A tool for identifying and addressing the root causes of critical incidents in healthcare.* Available at: www.patientssafetyinstitute.ca/resources/tools.html, retrieved November 10, 2008.

87. Institute for Safe Medication Practices Canada®. *Fluorouracil Incident Root Cause Analysis.* Web Format posting May 22, 2007. Available at: http://www.cancerboard.ab.ca/NR/rdonlyres/2FB61BC4-70CA-4E58-BDE-lEE54797BA47D/O/Fluorouracilincident/May2007.pdf, retrieved November 10, 2008.

88. Wikipedia, the Free Encyclopedia. *Thought experiment.* Available at: http://en.wikipedia.org/wiki/Thought_experiment>, retrieved November 20, 2008.

89. Holinagel E, Amalberti R. *The Emperor's New Clothes or whatever happened to 'human error'?* Invited keynote presentation at the 4th International Workshop on Human Error, Safety and System Development, Linkoping, June 11–12, 2001. Available also as Holinagel E. *The Elusiveness of "Human Error,"* Available at: http://www.ida.liu.se/~eriho/Human.Error_Mhtm, retrieved December 20, 2008.

90. Dörner D. The logic of failure. *Phil Trans R Soc Lond,* 1990; B **327**:463–73.

Chapter 3

Error and Harm in Health Sciences: Defining and Classifying Human Error and Its Consequences in Clinical and Community Settings

Executive Summary

The history of modern medical lathology began about 40 years ago. Most initial contributions related to medical error came from psychologists and specialists in cognitive science, studying both system and individual-, operator-, and human-produced error. Since the nineties, health professionals and new agencies (originally in the United States, Australia, the United Kingdom, and Canada) have transferred, adapted, and enriched nonmedical experiences and methodological developments, bringing them into the domain of healthcare including hospital and primary care, community medicine, and public health. From currently prevalent paradigms of system, or "no blame" error and error-producing individuals positioned as decision makers and executants of actions at the end of the chain of other contributing factors, comes a diverse and heterogeneous array of basic notions, terms, and definitions that are in part responsible for the still heterogeneous domain of medical or nonmedical lathology. Statisticians, informaticians, cognitive scientists, physicians, and nurses may use quite different definitions starting with error itself.

Definitions of error in a general and medical context, related events, possible causes, harm, and consequences are derived not only from their general or lexical meaning but also from the contributions of various domains and man- or machine-built specialties, organization, and functioning. Numerous heterogeneous meanings continue to prevail in medical error-related taxonomy.

From the examples of several definitions of terms and taxonomies with or without the semantic and lexical support described here arises a picture of various current taxonomies covering general and medical error domains. These taxonomies are consequently equally heterogeneous, and current efforts are directed, among others, to the development of some kind of general taxonomy that may or may not be suitable for various specific studies and health programs to explain, understand, prevent, and control medical error.

Let us adapt and adopt definitions that are not overlapping and redundant—operational enough for various specialists, actors, and environments in which medical error occurs—that allow integration and systematic reviews of various studies and other experience and some generalization to a specified degree.

Taxonomies themselves should permit systematizing errors and their types (as units of observation and analysis) as well as their potential and real causes and consequences including harm. They should also include other variables and phenomena of an administrative, managerial, technological, economical, social, and cultural and other relevant nature from one health problem to another.

Good definitions of terms and their taxonomies have and will have a profound influence on the value of descriptions, etiological research, medical error interventions effectiveness, and most strategies for the future.

In the next chapter, we will cover the challenges of describing medical error in general and some basic contributing methodologies to the study of medical error occurrence.

Thoughts to Think About

I see no dignity in persevering in error.

—**Robert Peel, 1833**

Error is not a fault of our knowledge, but a mistake of our judgment giving assent to that which is not true.

—John Locke, 1632–1704

The master of Trinity is correct: not one of us is infallible, not even the youngest ones.

—J. Chalmers Da Costa, 1863–1933

The medical errors of one century constitute the popular faith of the next.

—Alonzo Clark, 1807–1887

The physician in training who never makes a mistake takes his orders from one who does.

—Paraphrasing Lawrence J. Peter, 1977

Errors are not a "bad apple" problem where a handful of doctors or other medical personnel are the culprits and need to be rooted out or disciplined. Rather it is a systemic problem, where healthcare systems actually produce conditions that lead people to make mistakes or fail to prevent them. This means that we need rigorous changes throughout entire healthcare system to make it harder for people to do something wrong and easier for them to do things right.

—Donald Berwick, 2005

Individual human error should also be taken into account!

Medical error: hard to define; hard to find; hard to admit; hard to explain; hard to fix.

Introductory Comments

Medical error not only falls into the broader framework of human error as outlined in Chapter 2, but it also carries additional specific characteristics, requirements, objectives, and methodology particular to the health domain. How can we define it and understand its characteristics, requirements, and objectives? What is the basic model that would allow us to understand the methodology of its description, understanding, control, and prevention? What should physicians know and do? What should other health professionals and stakeholders know and do? How can patients and the community contribute through their active involvement in the best management of the medical error problem? This chapter should provide at least some essential answers to such essential questions.

It all starts with the current dizzying array of semantics and taxonomy in general and medical lathology. How can we see and understand this today?

Overview of Our Understanding of Error Today

Our current handling of errors in medicine builds on and benefits greatly from the experience of handling errors in daily life and various nonmedical domains of human activities as summarized in the previous chapter. Many elements (e.g., error terminology, definitions, taxonomy, methods and techniques of inquiry) of pioneering work by Joseph Jastrow,[1] Jens Rasmussen,[2] James Reason,[3] Carlo Cacciabue,[4] and others quoted in this chapter have been adopted, and, as we will see, have been adapted to the health sciences domain, be it biomedicine (as some alternative medicines protagonists call mainstream or "allopathic" medicine), some alternative and complementary medicines, nursing, nutrition, and other professions involved in healthcare. These years of work and development are now coupled, as they must be, with our own experience in fundamental, clinical, and epidemiological research, evidence-based medicine, and uses of evidence in structured critical thinking and decision making in patient care, community medicine, and public health. Invaluable contributions by psychologists and information technology scientists have set human error on a solid conceptual and methodological basis making one aspect of medicine easier and another more challenging by placing current experience in an even broader and deeper biological context and within the specific nature of human pathology and medical care. Chapter 2 reminded us that, in the past three generations, important contributions to the understanding and methodological development of the general and medical error domain were followed by an increasing number of key studies enriching practical experience with the reality of errors.

In 1936, Jastrow[1] and his group of representatives from numerous domains including the physical, biological, and human fields examined the omnipresence of error in human experience. Physiology, neurophysiology, anthropology, psychology, sociology, medicine (H. W. Haggard), and psychiatry (A. Myerson) were part of this vast coverage paving the way for the error concept and experience.

In 1987, Rasmussen et al.[2] applied the problem of error to the domain of new technology, industry, and various related occupations. In 1990, Reason[3] published a reference book about human error presenting among others the fundamentals of error definitions, taxonomy, and methodology based on the author's roots in psychology.

Shortly thereafter, John Senders and Neville Moray[5] provided an overview of the error problem based on gatherings of experts under the auspices of the U.S. Army, the North Atlantic Treaty Organization (NATO), and the Rockefeller Foundation. In 1994, Marilyn Sue Bogner[6] and her group of medical and non-medical experts translated the first error experience into the domain of medicine and built a bridge between the general error domain and the domain of medicine. More recently, in 2006, George Peters and Barbara Peters[7] also discussed human error, but from cause–effect and control perspectives.

Overview of Approaches to Error in Medicine

Remarkable brain trusts from various national and international health institutions were constituted to trace the present state of human and other error in medicine and to propose the development of still missing priorities in the advancement of our understanding of the error problem as well as error prevention and control. At least three major studies have demonstrated the seriousness of the error problem in medicine: one in Massachusetts (1991),[8,9] one in Australia (1995),[10,11] and one in Utah and Colorado (2000).[12] The Committee on Quality of Health Care in America in 2000 presented a document titled *To Err Is Human: Building a Safer Health System,*[13] which stresses the magnitude and seriousness of the medical error problem and the need for better knowledge through error reporting and for building performance standards and better safety systems in medical care.

Central governmental and other agencies like the Agency for Health Care Research and Quality (AHRQ), the Institute for Healthcare Improvement, the Joint Commission on Accreditation of Healthcare Organizations (JCAHO), the Canadian Institute for Health Information (CIHI), and the Canadian Patient Safety Institute (CPSI) deal entirely or partly with errors in clinical and community medicine care and their prevention and control, particularly in relation to patient safety and harm minimization. In 2009, the error experience was brought by Pat Croskerry et al.[14] into the field of emergency medicine in even greater and richer detail and applied to many other medical specialties as well. Individual, team, and system errors were in focus.

Not even a decade has elapsed since those basic contributions and experiences. This allows us to consider the following fundamental picture of human error in medicine and place it in the context of epidemiology, evidence, and evidence-based medicine and their uses in critical thinking and decision making in research and practice.

In "classical" or general lathology explained in the preceding chapter, events are the major subject of attention. Some of its consequences, like a broken device or the overall impact of error, often come only second. It is principally the study of events or happenings. In "medical" lathology, increased attention is paid to patients, their disease, and the disease outcomes and consequences as opposed to the events that lead to them. However, don't curing and caring about the patient also occur when we care and cure the avoidable harm that led to an additional health problem?

There are essentially two distinct domains of interest for definitions and their classification:

1. The error itself and other entities closely related to it.
2. Variables and phenomena related to error like risk characteristics, potential and real causes, as well as consequences of error.

Let us discuss them in this order.

Definitions of Medical Error, Associated Entities, Terms

Medical error is an error imputable to a physician as a health professional, the physician's teamwork and interaction with other health professionals, technology in clinical and community medicine care, the patient, and the physical and social environment of medical practice and research. Its various definitions reflect both general notions of error as reviewed in Chapter 2 and additional specific characteristics for the context in which medical errors happen. The domain also encompasses various natures of medical error, its causes, and occurrence. The frequency of medical errors differs from unique or sporadic cases such as failures of a surgical repair of a newly discovered vascular anomaly to more frequent errors like those underlying injuries inflicted to and by health professionals themselves including needle-prick injuries or medical practice acquired infections.

Ethical and social considerations sometimes lead to very broad paradigms and definitions of medical error, many of them excellent in concept and for motivational purposes and fewer of them being operational enough to make an epidemiologist happy in his or her research, intervention, and evaluation in the medical error domain. From this array of error paradigms, the core concept of error is twofold:

- Faulty reasoning, understanding, and ensuing decision making about a task or what to do.
- Sensory-motor faults in the execution of both the right and wrong task.

One definition of error in general was adopted by Croskerry et al.[14] in the domain of patient safety:

> An act of commission or omission that leads to an undesirable outcome or significant potential for such outcome: the process by which planned actions fail to achieve their desired ends. Three types of error fall into this same category: slips (attentional or perceptual failures), lapses (failures of memory) and mistakes (failures of mental processes).

As for medical error itself and associated terms, more specific definitions abound in health sciences, as follows.

Current Definitions of Medical Error and Medical Harm

Confusion still exists between understanding and defining *error as a failure in performing an action* and *the harm it produces*. Failures in performing an action may have as a cause some kind of "primary" error or failure in reasoning and decision making leading to a picture of errors as causes of other errors and so on. Very often, we still must read between the lines to distinguish causes of error, error itself, and the harm to which it may lead. For example, Newman-Toker and Pronovost stress this problem in the domain of diagnosis, but it extends across all elements and steps in medical care.[15]

Definitions of *medical error* span from general to some more operational ones:

- Anything that happened in your own practice that should not have happened, that was not anticipated and that makes you say "that should not happen in my practice, and I don't want it to happen again." It can be small or large, administrative or clinical—anything you identify as something to be avoided in the future.[16]
- Events in your practice that made you conclude, "That was a threat to patient well-being and should not have happened. I do not want it to happen again. Such an event affects or could affect the quality of the care you give your patients. Errors might be large or small, administrative or clinical, or actions taken or not taken. Errors might or might not have

discernible effects. Errors in this study are anything you identify as something wrong, to be avoided in the future.[17,18]

■ An inaccurate or incomplete diagnosis or treatment of a disease; injury; syndrome; behaviour; infection, or other treatment.[19]

■ The failure of a planned action in healthcare to be completed as intended or the use of an incorrect plan to achieve an aim.... Among the problems that commonly occur during the course of providing healthcare are adverse drug events and improper transfusions, surgical injuries and wrong-site surgery, suicides, restraint-related injuries or death, falls, burns, pressure ulcers, and mistaken patient identities. High error rates with serious consequences are most likely to occur in intensive care units, operating rooms and emergency departments.[13]

Definitions of *medical harm* cover either physical impairment only or also stress its relation to medical error:

■ A temporary or permanent physical impairment in body functions (e.g., sensory functions, pain, disease, injury, disability, death) and structures as well as suffering and other deleterious effects due to a disruption of the patient's mental and social well-being.[20-23]

■ Unintended physical injury resulting from or contributed to by medical care (e.g., the absence of indicated medical treatment) that requires additional monitoring, treatment, or hospitalization, or that results in death. Such injury is considered medical harm whether it is considered preventable, whether it resulted from a medical error, and whether it occurred within the hospital.... Some errors do indeed result in medical harm, but many errors do not; conversely, many incidents of medical harm are not the result of any errors.[24]

The latter appears more precisional than the former.

Let us note that medical harm is what ends up as subject of tort litigations at courts of law—not medical error, for which health professionals may be just disciplined. Chapter 8 discusses this in further detail.

Associated Entities, Terms, and Their Definitions

A comparison of the following definitions and taxonomies with those in the general error and harm domain as outlined in previous chapter will quickly show that the two do not often corroborate. Given the biological and other nature of health problems and their environment, adaptations to the health sciences domain appeared necessary and were and are made. For our purpose and understanding, a glossary at the end of this book somehow integrates both error and harm domains.

Why are we paying so much attention to definitions in this book?

Establishing rules, terms, and definitions is not only the starting point of any Socratic dialogue and discourse but is also of great importance in research and practice considerations including those focusing on medical error and harm. Research based on proper definitions should produce comparable and reproducible results representative of the problem. Without these results we can demonstrate and do nothing, despite any methodological fireworks that might ensue.

For Lavery and Hughes,[25] definitions fall into seven categories:

- Stipulative (those that coin new phenomena).
- Lexical or reportive (how the word is actually used).
- Precising (eliminating vagueness).
- Theoretical (formulating an adequate characterization).
- Motivational and persuasive (influencing attitudes in a metaphysical manner).
- Operational (those usable as a decision-making tool).
- Essentialist (specifying the nature of the phenomenon to which the term refers).

In our context, we need definitions that are first and foremost precise, theoretical, and operational. Without such qualities, any further uses of our scientific and practical experience will likely prove futile.

For pragmatic and practical purposes, definitions must allow for the following:[26]

- Measurement (*"That much?"* as in "How severe is the harm?").
- Counting (*"That many?"* as in establishing the frequency and distribution of errors).
- Classification and categorization (*"Where does all this belong?"* as in being usable in any meaningful taxonomy; the latter depending on the former).
- Being built with as exacting as possible inclusion and exclusion criteria (*"What's in?"* and *"What's out?"* questions further qualify in more operational terms the topic and its applications and uses in research and practice interpretations and decisions).
- Normal or abnormal qualification and decision power, usually combining frequency and biological plausibility considerations (*"Is it normal but perhaps too frequent?"* or *"Is it really abnormal whatever its frequency might be?"*).
- Evaluation (*"How was it?"* as in attributing the quality, relevance, and usability of findings).

If we look at definitions in this book and in the overall experience in general and medical lathology, the heterogeneity of the type of definitions and their required quality remain considerable. In the absence of a systematic evaluation, qualification, and categorization, the reader is relegated to perform this kind of exercise in the case of any entity of interest for his or her own work.

Two sets of entities are worth mentioning: (1) the error itself; and (2) other entities closely related to error that often carry overlapping definitions and meanings. The heterogeneity of many terms and their definitions still persists across the literature. Their following overview reflects such variety. The glossary at the end of this book offers their more unified view.

Definitions, examples, and the resulting harm that follow are based on Reason's[2] and Ferner and Aronson's[27] concepts and applications (corroborating largely with Reason[2]) and other references as specified under each item.

Human error: Any expected or unexpected human behaviour or lack of it leading to the undesirable consequences of an accident (vide infra); human involvement in accident causation. Human error produces events of the second order like an accident and its consequences (outcomes, harm). Creators and operators (vide infra) may both be responsible for human error.

Human error in medicine (our definition): A flaw in reasoning, understanding, and decision making made by a creator or operator regarding the solution of a health problem or in the ensuing sensory and physical execution of a task in clinical or community care. It is a product of the operator's body or mind dysfunction. An *operator in medicine* may be viewed in this context like a driver in the transportation industry or a machine operator in manufacturing. Physicians as decision makers who use diagnostic tests and devices (e.g., fiber optics, imaging techniques, psychiatric evaluations) or therapeutic tools (e.g., scalpel, laser, psychotherapy, electroshock, psychoactive drugs) are all "operators" in the process and system of medical care. The error is committed at the level of reasoning, critical thinking, and decision making or at the level of sensory or motor execution of the decided task and action and their evaluation. This definition fits into the framework of the aforementioned general definition of error as a failure of a planned action or use of an incorrect plan. This also includes the failure of a planned action to be completed as intended (i.e., error of execution) or the use of the wrong plan to achieve an aim (i.e., error of planning). Errors may be errors of commission or omission; they usually reflect deficiencies in the system of care.[28]

Knowledge-based error: Self-defined. *Example of error*: Being ignorant of the sciatic nerve anatomy. *Ensuing harm (consequence)*: Sciatic nerve palsy due to intramuscular injection.[27]

Rule-based error: (1) Misapplication of a good rule. *Example of error*: Prescribing oral treatment in a patient with dysphagia. *Consequence (resulting harm)*: Aspiration or failure to treat.[27] (2) A bad rule or failing to apply a good rule. *Example of error*: Prolonging antibacterial treatment beyond that which is necessary. *Consequence (resulting harm)*: Increased bacterial resistance.[27]

Action-based error (slip): Attentional or perceptual failure of an action.[27] *Example of error (slip):* Dispensing an irrationally high quantity of an anti-neoplastic drug. *Consequence (resulting harm):* Death.

Technical slip: Incorrect writing of orders, dispensing technique, inappropriate clinical maneuver execution. *Example of error (slip):* Giving intravenous injection by another way of administration. *Consequence (resulting harm):* Pain, loss of efficacy and effectiveness, tissue damage.

Memory-based error (lapse): Memory failure generated. *Example of error (lapse):* Forgetting a patient's allergy to an antibiotic. *Consequence (resulting harm):* Anaphylaxis.[27]

Medication error: Any preventable event that may cause or lead to inappropriate medication use or patient harm while the medication is controlled by the health professional, patient, or consumer. Such events may be related to professional practice, healthcare products, procedures, and systems, including prescribing; order communication; product labelling, packaging, and nomenclature; compounding; dispensing; distribution; administration; education; monitoring; and use.[29]

System error in medicine: Error imputable to technology and the environment of medical care and its interaction with users (health professionals as operators of the system) and recipients (patients and other receivers of preventive care).

Event: Any deviation from usual medical care that causes an injury to the patient or poses a risk of harm. Includes errors, preventable adverse events, and hazards.[28] Also includes something that happens to or with a patient or health professional or any other person involved in healthcare.

Adverse event: (1) Untoward incident, therapeutic misadventure, iatrogenic injury, or other adverse occurrence directly associated with care or services provided within the jurisdiction of a medical center, outpatient clinic or other facility.... Adverse effects may result from acts of commission or omission (e.g., administration of incorrect medication, failure to make a timely diagnosis or institute the appropriate therapeutic intervention, adverse reactions or negative outcomes of treatment).[30] (2) An unexpected and undesired incident directly associated with the care or services provided to the patient.[31] (3) An incident that occurs during the process of providing healthcare and results in patient injury or death.[31] (4) An adverse outcome for a patient, including an injury or complication.[31] (5) An injury related to medical management, in contrast to complication of disease. Medical management includes all aspects of care, including diagnosis and treatment, failure to diagnose or treat, and the systems and equipment used to deliver care. Adverse events may be preventable or nonpreventable.[28]

Accident: A specific, identifiable, unexpected, unusual and unintended event that occurs in a particular time and place of medical care, without originally apparent or deliberate cause, but with marked effects. In our context, accidents are the product of medical error.

Incidents: (1) Events, processes, practices, or outcomes that are noteworthy by virtue of the hazard it creates, or the harm it causes during the process of medical care related to the patient, health professional, and involved activities and tools. Despite a possible medical error at its origin, an incident does not necessarily produce an accident with its undesirable outcomes. An event that, under slightly different circumstances, could have been an accident.[31-33] (2) Patient safety events including adverse events, critical incidents, sentinel events, and near misses. "Incident" means any one of them.[34] (3) An event or circumstance that could have resulted, or did result, in unintended or unnecessary harm to a person or a complaint, loss, or damage.[20] (4) Any deviation from usual medical care that causes an injury to the patient or poses a risk of harm. Includes errors, preventable adverse events, and hazards.[28]

Critical incident: (1) An occurrence that might have led (or did lead)—if not discovered in time—to an undesirable outcome…. Complications that occur despite normal management are not critical incidents.[35] An event that involves no loss, but with a potential of loss under different circumstances. Other definitions across the literature and human experience abound (*vide infra*). (2) An incident resulting in serious harm (loss of life, limb, or vital organ) to the patient, or the significant risk thereof. Incidents are considered critical when there is an evident need for immediate investigation and response.[31,34] Outcomes that are unexpected in standard medical care and in the prevalent prognosis of the case.

Noncritical incident: Any undesired outcome-free incident. Readers and users should be aware of these different definitions. Incident definitions are among the most heterogeneous in this type of literature.

Near miss (close call, potential adverse event): (1) Any event that could have had adverse consequences, but did not and was indistinguishable from full-fledged adverse events in all, but outcome. An incident that did not cause harm.[20] (2) An undesirable outcome-free incident. (3) Serious error or mishap that has a potential to cause an adverse event but fails to do so because of chance or because it is intercepted.[28]

Sentinel event: (1) An unexpected occurrence involving death or serious physical or psychological injury or the risk thereof. Serious injury specifically includes loss of limb or function. The phrase "or the risk thereof" includes any process variation for which a recurrence would carry a significant chance of a serious adverse outcome. Such events are called "sentinel" because they signal the need for immediate investigation and response.[36,37] (2) Unexpected incident related to system or process deficiencies or human error that leads to death or major and enduring loss of function for a recipient of healthcare services.[34]

Root cause: Causal factors that, if corrected, would prevent recurrence of the incident, encompassing and derived from several contributing causes such as system deficiencies, management failures, performance errors, and inadequate organizational communication,[38] technology breakdowns, interfering physiological and psychological states of executing health professionals, their sensory-motor failures, or reasoning errors.

Mistakes: (1) Inappropriate planning (action specification) of medical care or research. Correct execution of an incorrect action sequence.[2,38] (2) Incorrect choices. Example: choosing the wrong diagnostic test or ordering a suboptimal medication for a given condition.[39] Reducing mistakes relies on improvements in training and supervision.[39] (3) Errors in planning of actions (knowledge-based or rule-based).

Slips: (1) Inappropriate actions and executions. Incorrect execution of a correct action sequence.[2,38] (2) Failures of schematic behavior or lapses in concentration, such as overlooking a step in a routine task due to a lapse in memory. Example: An experienced surgeon nicking an adjacent organ during an operation due to a momentary lapse in concentration.[39] (3) Failures due to competing sensory or emotional distractions, fatigue, or stress.[31] Reducing the risk of slips means improving the designs of protocols, devices, and work environments, reducing fatigue among staff, improving the design of key devices, eliminating distractions (phones), and other redesign strategies.[38]

Skill-based errors: Slips and lapses; errors in execution of correctly planned actions. They encompass both action-based errors (slips) and memory-based errors (lapses).[27]

Evidence-based errors: Errors due to a lack of the best evidence, the failure to use it, use of poor, unsupported, or inappropriate evidence in argumentations otherwise good or flawed, leading to incorrect claims and decision making in clinical practice and research. In the context of this writing, in agreement with Weston,[40] ignoring or misusing evidence and ignoring alternatives are the most significant fallacies as sources of error.

Argument-based errors: Misusing or omitting valid argument components or using them inappropriately (e.g., grounds, backing, warrants) and linking them poorly in medical decision making.

Failure: Nonperformance or inability of the system or component (human included) to perform an intended function in a specific persons/time/place context and conditions to reach an intended objective. Thus, not all faults are failures.

Latent failure (latent error): A defect in the design, organization, training, or maintenance in a system that leads to operator errors and whose effects are typically delayed.[28]

Error creators: Latent error-generating individuals at the "blunt end:" new medical technologies engineers, developers of clinical guidelines and work protocols, organizers of clinical care.

Operators: "Sharp-end," "acting" error makers like machine operators in manufacturing or drivers in the transportation industry. In health, they would include direct clinical care providers (prescribing physicians, operating surgeons, care-dispensing psychiatrists) using mental and sensorimotor-designed tools, machines, instruments, and materials (drugs, replacement substances).

Error outcomes: Incidents, accidents, and their physical, mental, and social consequences for the patient, health professional, and community. They may be negative (i.e., causing harm) or, less frequently, positive (changing health for the better).

Flaw: Given the often incriminating meaning of *fault, error, defect,* or *failure* assigned to these terms in the legal world as a possible cause of harm, some common term for possible (but not always necessary or unavoidable) causes of harm might prove useful. In this text, we will use the term *flaw,* generally defined as a feature marring the perfection of something or a defect impairing legal soundness or validity but used here as an encompassing denomination for ***any characteristic, deficiency, or inadequacy in technology, operation, human reasoning and decision making contrary to expectations from the expected function, task and the expected objectives and outcomes of a healthcare event.*** In other words, we propose flaw as a common term for any imperfection in evidence, reasoning, decision making, and consecutive sensory-motor action as well as in the evaluation of the results of such processes. Its possible incriminating quality must be proven; it is not a priori automatically incriminating itself. This term denotes any variable that may be considered as a possible (often to be yet proven) causal factor or marker of risk and prognosis in the domain of medical error and its consequences.

Safety: Freedom from accidental injuries.[28]

Patient safety: Reduction and mitigation of unsafe acts within the healthcare system as well as through the use of best practices shown to lead to optimal patient outcomes.[31]

Patient: Individuals, groups, populations, or entire communities who require care from healthcare professionals.[41]

System: A set of interdependent components (e.g., people, processes, equipment) interacting to achieve a common claim. System characteristics include complexity and coupling.[28,31]

Hazard: Any threat to safety (e.g., unsafe practices, conduct, equipment, labels, names).[28]

Critical Incident, Error, Harm: Comments on Current Terms Used in Medical Lathology

Lathologists, beware! Even if some terms are still subject to various definitions that are often too broad and sometimes still confusing, they reflect the increasing attention to and interest in the error problem in view of its better understanding and control.

As an example, let us examine here definitions of critical incident and the persisting confusion between error and its consequences (harm).

The term *critical incident* illustrates how definitions may still vary in the medical error domain. It is increasingly popular across written and electronic communication, sounds attractive, if not dramatic, for media and the public, but so far it may mean different things:

- Whatever is essential to change one's mind.
- Events significant or pivotal in some way or making a significant contribution to some events or phenomena.[39]
- One that makes a significant contribution—either positively or negatively—to an activity or phenomenon.[42]
- An incident not necessarily dramatic but that has significance for its perceiver, making him or her stop and think, raising questions, questioning one's beliefs, values, attitude, or related to and affecting communication, knowledge, treatment, culture, relationships, emotions, beliefs, or behavior and that has a significant impact on one's personal and professional learning either in a university or clinical setting.[43]

- Crucial information or event for a change of understanding or decision making.
- A stressful event like death, injury, or serious illness of someone in the work environment having an important, also qualified as traumatic, impact on workers' interaction and productivity and eventually on their own physical, mental, or social health.[42-46]
- A work-related traumatic event.[45]
- In work safety, then, events like serious, sudden, and unexpected injuries or deaths having significant psychological or other impact on employees are called critical incidents, requiring "critical incidence responses or techniques" to minimize their impact.[42]
- An unintended event that results from a patient's treatment in hospital and not from an underlying health condition.[46]

Moreover, work in the medical error domain is hampered by variations in the aforementioned entities from one study to another. For example, errors, accidents, incidents, failures, near misses, sentinel events, slips, and mistakes may still be used in research studies, as synonyms or not. The *Glossary of Terms* in the medical error and patient safety literature offered online by the US Agency for Healthcare Research and Quality (AHRQ)[39] and by the Royal College of Physicians and Surgeons of Canada[31] corroborates, however, in great part with several definitions previously quoted and is worthy of additional attention of the reader.

Specifying in each error research activity what will be the subject of the study is becoming mandatory. Chapter 2 and Appendices C and D of *To Err Is Human*[13] illustrate and justify better operational definitions that fit the objectives of each project. Even events that are subjects of surveillance in various states in the United States are, at times, accidents, incidents, error outcomes, or sentinel events.

In daily communication and other exchanges of professional and research information, occasional confusion between the terms *error* and *harm* may be noticed by readers and listeners as well. And so, looking once again at the previously defined definitions, we may correctly see that error is related to activities and events created in and by humans, teams, or systems. Harm means bodily, physical, mental, or social impact on patients as individuals, groups, and communities.

Given that error does not always end up in all cases as harms, epidemiologically speaking, registering and analyzing cases of harm means analyzing only the possible or real tip of the iceberg of an entire error problem reflected in the persons/time/place picture of harm. Potential overlaps must be considered and analyzed on a case-by-case basis.

Variables and Their Taxonomy in the Medical Error Domain

Taxonomy, the practice and science of classification, is particularly challenging in the medical error domain. It must cover not only various types of error but also their presumed and real causes and consequences, remote factors (latent causes), and proximate (patent, salient, apparent, direct, active, human, triggering) factors, be they from physical or human environments.

There is no universal taxonomy of medical error and taxonomies across the available experience around,[47] related and built to answer questions about a medical error to be solved: for example, errors in different environments of medical practice (e.g., hospital based, ambulatory care, medical specialties); errors in various domains of research; errors in various steps of clinical work like risk assessment, diagnosis, treatment, or prognosis; errors in clinical and community care. The same applies to the classification of variables or factors involved in the error study. The following (abridged for explicitness) taxonomies of error itself and variables and potential causes and other technical, administrative, process, and resources characteristics[15,32] may serve as examples. Their choice or modification depends on the question we ask in the study of medical error. There is no universal taxonomy of medical error, since the more it is general and encompassing, the less it is operational and usable in research and practice focusing on a specific problem and domain.

The *general purpose of any taxonomy* related to any human experience is to organize all available, necessary, and relevant information for a particular function and to provide a framework to answer a particular set of questions.[47]

The *purpose of taxonomy in the patient safety and medical error domain* is to organize the available and developing knowledge and information about patient safety for use in clinical practice and research and to provide a framework to describe the occurrence rates of specific types of events, the relationships between contributing factors, and the domains of medicine in which events occur.[47]

Exploring well-classified (i.e., "taxonomized") data should help to describe and explain what happened, in what context, the contributing factors to the event (or events), and the outcomes of such events including unwanted medical outcomes or harm. As generators of causal hypotheses, they should provide the basis for further analytical (etiological) research, most often beyond the scope of the taxonomy used. In this spirit and for such purposes, Chang et al.[48] expect an acceptable taxonomy to be based on an unambiguous terminology and classifications, useful for both analyses of risk and prognosis (together with root causes and contributing factors, allowing consistent collection and analysis across the continuum of healthcare settings including near misses and adverse effects), movement of information and findings, and setting priorities for interventions to improve patient safety. Ensuing corrective measures and their evaluation should not be forgotten either.

Figure 3.1 illustrates such multipurpose expectations from taxonomy of medical error.

The most challenging task is to provide some kind of universal or "all-purpose" taxonomy, still not within reach. The type of taxonomy will always depend on questions raised and answers to be given.

Often, taxonomies of "error" are taxonomies of variables possibly leading to error. They must be read and interpreted in this spirit. Ideally, models covering variables related to error should present a balanced view of both latent and active errors and their possible determinants (independent variables related to it).

Migration of Error Taxonomy from Industry to Health Sciences: An Example

One good example of medical error coverage may be the Eindhoven classification of error. First developed for the chemical industry[49] and later on adapted to medicine,[50] the list of error-related variables or events as summarized in Table 3.1 include errors and their possible determinants and other factors from both the latent and active domains.

This example still does not cover in a balanced way reasoning, critical thinking, and decision-making factors and errors within the active domain and between the latent and active domains as previously mentioned. We will attempt to establish this balance in Chapter 5.

Medical Error and Related Factors and Variables: Other Approaches

Other taxonomies are taxonomies not only of error itself but also of other variables, circumstances, and settings related to it. While the Linnaeus-PC Collaboration's taxonomy (in a later section in this chapter) combines types of error with circumstances, conditions, consequences, and corresponding corrections, Zhang et al.'s taxonomy[38] discussed herein pays special attention to slips and mistakes in that domain. This chapter will present an idea of taxonomy particularly linked to clinical factors and medical specialties and will give an example of an exhaustive taxonomy covering most of what has already been mentioned.

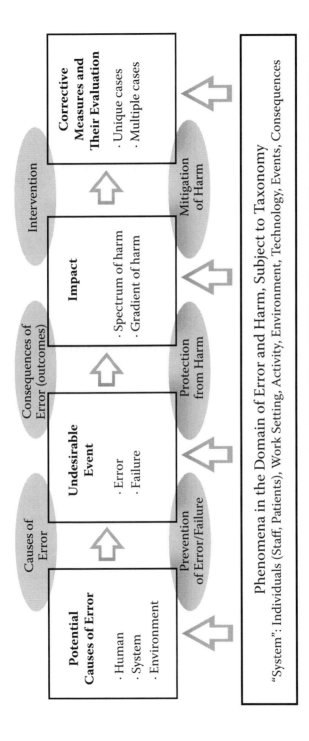

Figure 3.1 **What events and relationships between them should be covered by a taxonomy of error and harm.** (Modified and expanded from Chang, A. et al., *Int. J. Qual. Health Care*, 17, 2, 2005.[48])

Table 3.1 The Eindhoven Classification of Error and Its Determinants Applied to Medicine and Other Health Sciences[50]

A. Latent Domain	
1. Technical domain (e.g., physical environment, equipment, material, their content and design defects)	External factors and errors (technical failures or factors beyond the control of an interested party) Incorrect construction of a correct design of the above
2. Organizational domain	External (factors and failures beyond the control of an interested party) Transfer of knowledge (faulty training and communication between actors) Protocols/procedures (deficient guidelines) Management priorities (failures in decision making and relegating hierarchy creating incompatibility between internal management decisions and conflicting objectives and needs) Culture (failures from a prevailing collective approach, way of thinking and behavior to error and its determinants)
B. Active Domain (Human Behavior Errors and Their Determinants)	
1. External	Human failures beyond the control of an interested party
2. Knowledge-based	Misapplications of existing knowledge to both known and novel situations including *rule-based* behaviors (faulty qualifications of team members, lack of coordination between team members, deficient evaluation and ongoing verifications of the medical care process, faulty task planning and execution of intervention, monitoring the process and outcomes of interventions). N.B. Simply not following rules, guidelines, standards, or norms is often labeled a "cause" without formal causal proof.

Table 3.1   The Eindhoven Classification of Error and Its Determinants Applied to Medicine and Other Health Sciences[50] (*Continued*)

3. Skill-based	Behaviors such as slips (performance of fine motor skills) and tripping (inadequate whole body movements). Reminder: Skill-based active or human errors are, however, more complex than movement related errors. Sensory performance failures in which vision, auditory, tactile, olfactory, or taste functions may be involved as well as motor errors like posture and gross movements, fine "execution" of movements and ergonomic subjects like good or bad tool (instrument, machine, device) selection and uses, early failures to detect breakups or any combination of the above may be questioned.
C. Other	
1. Patient-related	Factors (patient characteristics, state, conditions, morbidity and comorbidity, values or actions) that influence clinical care and treatment beyond the control of staff
2. Unclassifiable	Factors that cannot be classified in any of the aforementioned categories

Source: Reworked from Battles, J. B. et al., *Arch Pathol & Lab Med,* 122, 3, 1998. With comments added.

Taxonomy by Types, Circumstances and Conditions, Consequences, and Corrections of Medical Error

As Table 3.2 shows, this taxonomy is built according to the following axes: the error itself; and related dependent and independent variables.

Slips and Mistake-Related Taxonomy

Zhang et al.[38] propose a cognitive taxonomy of errors based on their relationship with slips and mistakes and error-generating activities (executions and evaluations) as illustrated in Figure 3.2.

Table 3.2   Taxonomy by Types, Circumstances and Conditions, Consequences, and Corrections of Medical Error: The Linaeus-PC Collaboration's *International Taxonomy of Medical Errors in Primary Care—Version 2*[16]

Medical Error and Its Related Factors
The Error Itself

A. Error Types
 	
 1. Process errors
 	
   a. Office administration (chart completeness and availability, patient flow, message handling, appointments)
 	
   b. Equipment and physical building, surroundings, practice site (building infrastructure or management errors, practice errors in using equipment)
 	
   c. Errors in paraclinical investigations (laboratory and imaging tests including their subsequent use)
 	
   d. Treatment errors (ordering, receiving, following medications)
 	
   e. Communication between health professionals and patients
 	
   f. Payments (insurance and electronic handling related)
 	
   g. Workforce and organization of care (organization and management of staff and its workload)
 	
 2. Knowledge and skills errors
 	
   a. Execution of a clinical task
 	
   b. Execution of an administrative task

B. Related Factors as Dependent or Independent Variables   1. Actions taken (what a provider or practice did as a result of an event like harm control, at the clinical or paraclinical levels)   2. Consequences (error outcomes): patient, family, health provider   3. Harm (occurrence, degree)   4. Contributing factors   a. *Patient factor* (same or similar names, culture, health, behaviour, family knowledge and understanding, communication, other)

Continued

Table 3.2   Taxonomy by Types, Circumstances and Conditions, Consequences, and Corrections of Medical Error: The Linaeus-PC Collaboration's *International Taxonomy of Medical Errors in Primary Care—Version 2*[16]

b. *Provider factor* (same or similar names, culture, character and motions, health, behaviour, experience, communication)

c. *Provider team factor* (teamwork, experience, constitution, communication, other)

d. *Task factor* (characteristics, functioning, clinical characteristics, communication, other)

e. *Working conditions* (time, culture, equipment, distractions, communication, other)

f. Organization factor (care coordination, clinical, administrative dysfunctions, referral services, changes in systems, other)

g. *Physical environment* (timing of care, weather, location, proximity, appearance of services)

h. *Regulatory/payment system factor* (implementation of regulations, changes, regulatory and legal obligations, cost of care)

i. *Don't know*

C. Prevention Strategies

0. No prevention strategy offered

1. Avoidance of mistakes in doing the same

2. Providing and receiving a different care (patient and provider behavior, assessment, appointments, records and recording, clinical and paraclinical investigations, treatments, alerts, checking, workforce, workplace and equipment organization, management with other settings)

3. Better communication (teams, systems, technology)

4. Better education and training (patient, health professionals, or individuals at future risk)

5. More time or money

Source: Abridged and modified from The Linnaeus-PC Collaboration. *International Taxonomy of Medical Errors in Primary Care—Version 2.* Washington, DC: The Robert Graham Center, 2002.

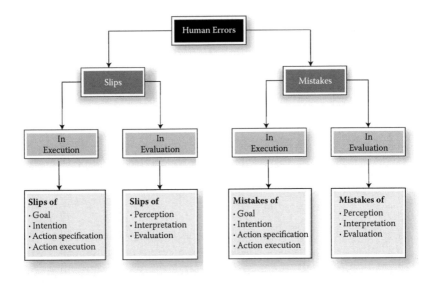

Figure 3.2 Human errors as slips or mistakes. (Redrawn and expanded with modifications from Zhang, J. et al., *J. Biomed. Informatics*, 37, 2004.)

Clinical Factors and Specialty-Oriented Taxonomies

Two examples may be quoted here: one from pediatrics and the other from clinical pharmacology.

1. Pediatrics: Woods et al.[47] proposed a list of variables of interest in a medical error domain taxonomy for pediatrics and easily modifiable and adaptable for a broader use across medicine. Their classification proceeds on the basis not of general experience in lathology but rather of what interests clinicians and what might be brought in to the larger context of medical error research and patient safety improvement. Briefly, *Pediatric Patient Safety Taxonomy*[47] includes the following:
    a. *Event types* (e.g., problematic decisions, executions, communications).
    b. *Domains of medical care* (preventive medicine like screening and immunization, steps in clinical work like diagnosis, treatment by surgical and nonsurgical procedures, patient monitoring, other services, communication, administration factors).
    c. *Contributing factors* (patient–child specific physical characteristics, physiological and cognitive development, medical and administrative management of the case including selected legal aspects, other noncontributing or information deficient factors, human factors like knowledge

and cognition, health professional's physical and emotional state, or latent conditions).

    d. *Outcomes of care* (near misses, adverse drug events, emergencies and critical clinical incidents beyond the natural history or course of the patient, as well as the level of harm to which they are associated).

2. Clinical pharmacology: In adverse drug events and medication errors, the taxonomy must encompass several axes specific to this specialty including the following as proposed by Morimoto et al.:[51]

    a. Sources of reporting.

    b. Symptoms and actions suggesting potential adverse drug events or medication errors.

    c. Diagnoses which follow adverse drug events.

    d. Type of drugs involved and potentially related to adverse drug events.

    e. Drug combinations.

    f. Clinical manifestations (signs and symptoms) potentially related to drug combinations.

    g. Miscellaneous factors related to the combinations of drugs (factors and markers).

    h. Laboratory results that trigger further inquiry.

    i. Type of incident and resulting potential error with or without harm.

    j. Severity of harm potentially related to the incident.

In ambulatory care, Gandhi et al.[52] note 27% of adverse drug events in patients surveyed, 13% of those being serious, 11% preventable, 28% ameliorable, and 6% either preventable or ameliorable.

Exhaustive and Multi-Axial Taxonomies

Any taxonomy is (or should) be built for use in specific research activities and healthcare improvement programs and evaluations. All depends then on their objectives. For this reason, some universal taxonomy and classification still remain a remote goal and objective.

Three examples of such valuable initiatives are worth mentioning: the ICPS (International Classification for Patient Safety) Study, the WAPS (World Alliance for Patient Safety) Study, and the TAPS (Threats to Australian Patient Safety) Study.

The ICPS Study

As a recent contribution to the multiple ways of classifying and defining[53] error and its causes, the World Health Organization's (WHO's) *International*

Classification for Patient Safety[22] (still in its testing stage) is one of the most complete. It also includes another glossary of preferred terms and definitions as well as a taxonomy that covers the following:

- *Incident type,* by its clinical administration, clinical process/procedure, documentation, healthcare associated infection, medication/IV fluids, blood/blood products, nutrition processes and problems, oxygen/gas/vapor problems, medical device/equipment/property uses and misuses, patient behavior, patient accidents (ways and types of exposure), infrastructure/buildings/fixtures list and quality, resources/organizational management type contributing to error, and pathology/laboratory step-by-step activities—sites of error.
- *Patient outcomes* by anatomical-pathological system and degree of harm (severity).
- *Patient characteristics* by demographics, type of care, and primary diagnosis.
- *Incident characteristics,* care setting of the incident, steps of inpatient management, specialties or disciplines involved in care, person reporting (healthcare professional, healthcare worker, community emergency services personnel, patient and close laics), people involved in care and incident management (healthcare professionals and others).
- *Contributing factors/hazards* (staff, patient, work/environment, organizational/service, external factors, other).
- *Staff-related contributing factors/hazards* (cognitive, performance, behavior/violation, communication, patho-physiologic/disease related, staff emotional and social factors).
- *Patient-related contributing factors/hazards* (same staff-related classification as previously mentioned).
- *Work/environment-related contributing factors/hazards* (physical, functional).
- *Organizational/service-related contributing factors/hazards.*
- *External factors-related contributing factors/hazards* (natural environment, products/technology/infrastructure, services/systems/policies).
- *Mitigating factors* from patient, staff, organization, or agent sides.
- *Detection* (by whom and by nature/type/mechanism of recognition).
- *Organizational outcomes* (property, patient, civic-life related).
- *Ameliorating actions* (patient related, organization related).
- *Actions to reduce risk* focusing on patients, staff, organization and environment, agent and equipment, and other.

The WAPS Study

At the time of this writing, *The World Alliance for Patient Safety Drafting Group* has further advanced work in international terminology and taxonomy in the medical error and harm domain.[21,53-55]

In this study methodology, causes (factors and markers of incidents) are retained (contributing factors and hazards, patient characteristics), incident types and characteristics, ways of detection, mitigating factors, and outcomes (patient, organizational) as well as ameliorating actions are examined. The system is then primarily process oriented.[54]

The TAPS Study

The *Threats to Australian Patient Safety (TAPS) Study*[56] and the ensuing proposed taxonomy includes, in an abridged form, a more extended array of safety events reported in general practice:

1. Errors related to the processes of healthcare.
    a. Errors in practice and healthcare systems.
    b. Investigation errors.
    c. Medication errors.
    d. Treatment errors (nonmedication).
    e. Communication errors and process errors not otherwise specified.
2. Errors related to the knowledge and skills of health professionals.
    a. Errors in diagnosis.
    b. Errors in managing patient care.

The authors hope that their proposal will remove difficulties arising from coder interpretation of events.[56]

Notes about Related Variables and Contributing and Mitigating Factors

The selection of the number of variables to study in the error domain does not always follow William of Occam's (1285–1349) "Razor" (i.e., saying that "assumptions to explain a phenomenon must not be multiplied beyond necessity"). This maxim is cherished by epidemiologists. However, difficult logistics in the study of error in the health domain, human and material resources requirements, and time frequently push investigators to gather as much information as possible when opportunity knocks. Relevant information must then often be selected a posteriori from large data banks. Also, definitions can be satisfactory to a variable degree from one study to another.

Note about Related Variables

Given the view of medical error as an individual-, system-, or environment-related problem, the list of variables related to error is almost endless and too extensive to be operational from one study to another. For example, almost 14 pages of the Linnaeus-PC Collaboration's *International Taxonomy of Medical Errors in Primary Care*[16] are devoted to a list of such variables. From this kind of a set, variables must be chosen selectively, be it for surveillance and general reporting or for the study of a more narrowly defined research or evaluation activity.

In addition to the extensive variables problem, researchers will seek in many cases usable definitions, ideally with inclusion and exclusion criteria, to know more precisely what or who is in and what or who isn't. For example, researchers will examine how to more precisely define *complex social situation, new staff, difficult diagnosis,* or *no continuity of care.*

Lists of variables related to the origin and causes of medical error or to its related consequences must be defined and developed from one study to another to suit the nature and objectives of the study and to allow comparisons of results of other studies of interest. Differences in definitions and taxonomy of any kind make comparisons of results more difficult, if not impossible. Any meaningful research synthesis and meta-analysis of findings may prove equally challenging.

On the other hand, more "complete" lists make us more aware of the complexity of the medical error problem and lead us to a better selection of variables of interest needed from one case to another.

Note about Contributing and Mitigating Factors

How can we understand and interpret the meaning of the contributing and mitigating role of some variables in the medical error domain? They are an example of epidemiological challenges that may be found in the error domain.

A *contributing factor* is defined as a circumstance, action, or influence (e.g., poor rostering or task allocation) thought to have played a part in the origin or development of an incident, or to increase the risk of an incident[39] (and the severity of outcomes). A contributing factor may be external, organizational, staff, or patient related. In fire protection,[57] contributing factors are any behavior, omission, or deficiency that sets the stage for an accident or increases the severity of injuries. This is opposed to "causal factors," defined in fire protection as any behavior, omission, or deficiency that, if corrected, eliminated, or avoided, probably would have prevented a fatality.

A *mitigating factor* is defined as an action or circumstance that prevents or moderates the progression of an incident toward harming a patient.[22] Mitigating factors have a different meaning in the legal domain.

Some epidemiologists may find it difficult to know if and how to place those factors within a web of incident causes and what their proportional contribution to the etiology of an incident may be. Given that their role in causal research is poorly known or unknown in quantitative and more operational terms, *potentially contributing* or *potentially mitigating* qualifications of factors might be more realistic adjectives.

Conclusions: Implications of Definitions and Taxonomy for Research and Management of the Medical Error Domain

Is medical error different from human error? Yes it is, in its premises and conclusions that help us survive or not. Definitions of observations in the medical error domain as well as their taxonomy are crucial for further research and understanding of the error problem. We want and prefer to adopt definitions that

- Are not redundant or overlapping.
- Are operational enough to allow them to be used the same way from one investigator (and reader) to another and from one site of research to another.
- That allow some degree of integration of various studies and their findings.
- That permit some degree of generalization when needed.

Many definitions still vary from one glossary to another. This diversity is amplified by the fact that authors of glossaries and other reference sources may be physicians, nurses, psychologists, specialists in operational research or informaticians, or various combinations of these. The meaning of definitions does not always migrate well from one domain of expertise to another. Moreover, what is operational for a psychologist or psychiatrist is not often operational enough for a biochemist, anatomist, surgeon, pharmacologist, or anthropometrist.

For now, let us always specify definitions of all variables discussed. In this book, we are using all terms as defined in the glossary at the end of the text. Lathology in medicine is about a quarter of a century young. It is still evolving.

Available taxonomies are often built for descriptive purposes in an attempt to create a picture of distribution of error occurrence in persons/time/place context and setting. For analytical purposes, taxonomies may include the following:

■ Errors and their type as a basic unit of observation and analysis.
■ Errors classified as consequences of something.
■ Errors classified as causes of something.
■ Potential causes (independent variables) that may lead to them.
■ A still poorly defined and delimited category of "contributing factors."
■ Various outcomes consequent to various types of error.
■ Other variables of administrative, managerial, economical, social, cultural, and different natures offering varying understanding of their purpose in the taxonomy.

Let us remind ourselves that any taxonomy is meaningful and usable only if it is supported by operational definitions of its components, terms, and entities. Two major challenges in dealing with medical error are the following:

1. Harm and error have a different meaning and are separate entities. Still, we must sometimes read between the lines to know if lathologists cover in their messages error, harm, or both.
2. In observational etiological inquiries and experimental research, trials and program evaluations at any level of prevention may both play the role of dependent or independent variables (among all other relevant variables).

Despite all initiatives described in this chapter, we will not one day necessarily have a universal, "all-purpose" taxonomy. We will sometimes, if not often, prefer a taxonomy "built to measure" that allows a search for answers and the reaching of research objectives from one study of medical error to another. An ever encompassing taxonomy containing "all" categories possible and all real and foreseeable dependent and independent variables is not necessarily the most operational and usable one.

Once hypotheses about the medical problem are raised, variables of interest defined and ways to classify and use them determined, we may proceed to the study and understanding of medical error occurrence (as outlined in Chapter 4), analyze their causes or consequences (Chapter 5), and attempt other programs of a more or less experimental (controlled) nature for even better understanding, prevention, and control. Only the future will show if some universal taxonomy in the medical error domain is possible. The author would be the first to be happy if proven wrong.

In principle, based on experience from other domains of epidemiology, the more tools and systems are exhaustive, complete, and encompassing, the less they are operational. All research studies and health programs have their own specific objectives, and taxonomies must satisfy such goals. On the other hand, definitions must be as explicit, exhaustive, mutually exclusive, and formulated not only in conceptual but mainly also in operational terms (ideally, with inclusion and exclusion criteria). Currently, definitions and glossaries in the medical error domain satisfy more, leaving taxonomies for further refinements to respond in the best way possible to research studies and health programs in prevention and to control goals.

Tackling the medical error and harm problem is challenging because it is based on fundamental and remarkable contributions from several domains more or less related (and also originally unrelated) to health sciences. Enter now clinical sciences, public health and community medicine, epidemiology, and biostatistics. A common platform is being built as this chapter illustrates, and it must be so. Here, and beyond the challenge of taxonomy, we believe that the nature and occurrence of medical error and harm do not make them a subject suitable for some sort of uniform methodology of their study and management. The domain of medical error and harm looks for now like a dual research and control challenge. From methodological and research strategy standpoints, medical error and harm fall into two main, distinct, and complementary categories that may require different and complementary methods:

- *Unique or rare cases and events.* The most in-depth original experience in lathology was acquired in this domain. Given that these events are another type of health problem such as disease cases, modern methodology of clinical case reporting, case study, qualitative inquiry, or clinimetrics may prove a useful addition to the methodological armamentarium initially developed by psychologists, operational research specialists, and other lathologists.
- *More frequent cases and events* require even more epidemiological involvement in addition to current experience acquired through other means of study and intervention. Series of cases of error and harm occur and must be subjected to epidemiological description, analysis, search for effective control, and evaluation.

This reality may also constitute a challenge in medical error and harm research synthesis, be it in the form of meta-analyses or systematic or other reviews. Strategies of intervention, health (medical error and harm) programs, and their evaluations must adapt to this kind of reality. The chapters that follow must be read and understood in this double spirit.

References

1. *The Study of Human Error.* Edited by J Jastrow. New York: D. Appleton-Century Company, 1936.
2. Reason J. *Human Error.* Cambridge, UK: Cambridge University Press, 1990.
3. *New Technology and Human Error.* Edited by J Rasmussen, K Duncan, J Leplat. Chichester, UK: John Wiley & Sons, 1987.
4. Cacciabue PC. *Guide to Applying Human Factors Methods. Human Error and Accident Management in Safety Critical Systems.* London: Springer Verlag, 2004.
5. Senders JW, Moray NP. *Human Error: Cause, Prediction, and Reduction. Analysis and Synthesis.* Hillsdale, NJ: Lawrence Erlbaum Associates, Publishers, 1991.
6. *Human Error in Medicine.* Edited by MS Bogner. Hillsdale, NJ: Lawrence Erlbaum Associates, Publishers, 1994.
7. Peters GA, Peters BJ. *Human Error. Causes and Control.* Boca Raton, FL: CRC/ Taylor & Francis Group, 2006.
8. Brennan TA, Leape LL, Laird NM, Hebert L, Localio AR, Lawthers AG, et al. Incidence of adverse events and negligence in hospitalized patients: results of the Harvard Medical Practice Study I. *N Engl J Med*, 1991;**324**:370–6.
9. Leape LL, Brennan TA, Laird N, Lawthers AG, Localio AR, Barnes BA, et al. The nature of adverse events in hospitalized patients: results of the Harvard Medical Practice Study II. *N Engl J Med*, 1991;**324**:377–84.
10. Wilson RM, Runciman WB, Gibberd RW, Harrison BT, Newby L, Hamilton JD. The Quality in Australian Health Care Study. *Med J Austr*, 1995;**163**:458–71.
11. Wilson RM, Harrison GB, Gibberd RW, Hamilton JD. An analysis of the causes of adverse events from the Quality in Australian Health Care Study. *Med J Austr*, 1999;**170**:411–5.
12. Thomas EJ, Studdert DM, Burstin HR, Orav EJ, Zeena T, Williams EJ, et al. Incidence and types of adverse events and negligent care in Utah and Colorado. *Med Care,*1999;**38**(3):261–71.
13. Institute of Medicine, Committee on Quality of Health Care in America. *To Err Is Human: Building a Safer Health System.* Edited by LT Kohn, JM Corrigan, MS Donaldson. Washington, DC: National Academy Press, 2000.
14. Croskerry P, Cosby KS, Schenkel SM, Wears RL. *Patient Safety in Emergency Medicine.* Philadelphia: Wolters Kluwer | Lippincott Williams & Wilkins, 2009.
15. Newman-Toker DE, Pronovost PJ. Diagnostic errors—the next frontier for patient safety. *JAMA,*2009;**301**(10):1060–2.
16. The Linnaeus-PC Collaboration. *International Taxonomy of Medical Errors in Primary Care—Version 2.* Washington, DC: The Robert Graham Center, 2002. (Last updated August 30, 2004.)
17. Rosser W, Dovey S, Bordman R, White D, Crighton E, Drummond N. Medical errors in primary care. Results of an international study of family practice. *Can Fam Physician*, 2005;**51**(3):386.
18. Jacobs S, O'Beirne M, Derflinger LP, Vlach L, Rosser W, Drummond N. Errors and adverse events in family medicine. Developing and validating a Canadian taxonomy of errors. *Can Fam Physician*, 2007;**53**:270–6.

19. Wikipedia, the Free Encyclopedia. *Medical error.* Available at: http://en.wikipedia.org/wiki/Medical_error, retrieved December 23, 2008.

20. Runciman WB. Shared meanings: preferred terms and definitions for safety and quality concepts. *MJA,* 2006;**184**(10):S41–S43.

21. Runciman W, Hebert P, Thomson R, van der Schaaf T, Sherman H, Lewalle P. Towards an International Classification for Patient Safety: key concepts and terms. *Int J Qual Health Care,* 2009;**21**(1):18–26.

22. WHO World Alliance for Patient Safety. *The Conceptual Framework for the International Classification for Patient Safety (ICPS). Version 1.0 for Use in Field Testing 2007–2008.* Geneva: World Health Organization, July 2007. Available at: http://www.who.int/patientsafety/taxonomy/icps_download/en/print.html, retrieved February 4, 2009.

23. Institute for Safe Medication Practices Canada; c2000-2006. *Definitions of Terms.* Available at: http://www.ismp-canada.org/definitions.htm, retrieved July 9, 2009.

24. Gold JA. The 5 Million Lives Campaign: Preventing medical harm in Wisconsin and the nation. *Wisc Med J,* 2008;**107**(5):270–1.

25. *Meaning and Definition.* Chapter 2, pp. 33–60 in: *Critical Thinking: An Introduction to Basic Skills.* Fourth Edition. Edited by J Lavery, W Hughes. Peterborough, ON: Broadview Press Ltd., 2004.

26. Jenicek M. Four cornerstones of a research project: Health problem in focus, objectives, hypothesis, research question. Chapter 3 (pp. 27–34) in: *Biomedical Research: From Ideation to Publication.* Edited by G Jagadeesh, S Murthy, YK Gupta, A Prakash. New Delhi and Philadelphia: Wolters Kluwer Health/Lippincott/Williams & Wilkins, 2010.

27. Ferner RE, Aronson JK. Clarification of terminology in medication errors. Definitions and classification. *Drug Safety,* 2006;**29**(1):1011–22.

28. WHO, World Alliance for Patient Safety. *WHO Draft Guidelines for Adverse Event Reporting and Learning Systems. From Information to Action.* Geneva: World Health Organization (document WHO/EIP/SPO/QPS/05.3), 2005.

29. National Coordinating Council for Medication Error Reporting and Prevention. *What Is a Medication Error?* Available at: http://www.nccmerp.org/aboutMedErrors.html?USP_Print=true&frame=lowerfrm, retrieved January 14, 2009.

30. US Department of Veteran Affairs. National Center for Patient Safety. *Glossary of Patient Safety Terms.* Available at: http://www.va.gov/NCPS/glossary.html, retrieved November 9, 2008.

31. Davies JM, Hébert P, Hoffman C (The Royal College of Physicians and Surgeons of Canada). *The Canadian Patient Safety Dictionary.* Ottawa: Royal College of Physicians and Surgeons of Canada, October 2003. Available at: http://rcpsc.medical.org/publications/PatientSafetyDictionary_e.pdf, retrieved January 21, 2009.

32. Barach P, Small SD. Reporting and preventing medical mishaps: lessons from non-medical near miss reporting systems. *BMJ,* 2000;**320**:759–63.

33. National Research Council, Assembly for Engineering, Committee on Flight Airworthiness Certification procedures. *Improving aircraft safety: FAA Certification of Commercial Passenger Aircraft.* Washington, DC: National Academy of Sciences, 1980.

34. Baker R, Grosso F, Heinz C, Sharpe G, Beardwood J, Fabiano D, et al. Review of provincial, territorial and federal legislation and policy related to the reporting and review of adverse events in healthcare in Canada. Pp. 133–137 in: *Adverse Event Reporting and Learning Systems: A Review of the Relevant Literature.* Edited by JL White. Edmonton: Canadian Patient Safety Institute, June 25, 2007. Available at: www.patientsafetyinstitute.ca/uploadedFiles/News/CAERLS_Consultation_Paper_AppendixA.pdf, retrieved January 26, 2009.

35. Busse DK. *Cognitive Error Analysis in Accident and Incident Investigation in Safety-Critical Domains.* A PhD Thesis submitted at the University of Glasgow Department of Computing Science. Glasgow: University of Glasgow, September 2002. Available at: http://www.dcs.gla.ac.uk/~juhnso/papers/Phd_DBusse.pdf, retrieved October 25, 2008.

36. The Joint Commission. *Sentinel Event Glossary of Terms.* Available at: http://www.jointcommission.org/SentinelEvents/se_glossary.htm, retrieved December 29, 2008.

37. The Joint Commission. *Annotated Resources—Sentinel Events.* Available at: http://www.jointcommission.org/SentinelEvents/se_annotated.htm, retrieved December 29, 2008.

38. Zhang J, Patel VL, Johnson TR, Shortliffe EH. A cognitive taxonomy of medical errors. *J Biomed Informatics,* 2004;**37**:193–204.

39. US Agency for Healthcare Research and Quality (AHRQ). *Glossary.* Available at: http://www.wbmm.ahrq.gov/popup_glossary.aspx, retrieved January 29, 2009.

40. Weston A. *A Rulebook for Arguments.* Third Edition. Indianapolis: Hackett Publishing Company, 2000.

41. Wikipedia, the Free Encyclopedia. *Critical Incident Technique.* Available at: http://en.wikipedia.org/wiki/Critical_Incident_Technique, retrieved January 8, 2009.

42. Monash University. *What is a 'critical incident'?* Available at: http://www.monash.edu.au/lls/llonline/writing/medicine/reflective/2.xml, retrieved January 8, 2009.

43. Workplace Safety & Insurance Board Ontario (WSIB), OW Safety–2005. *Handling a critical incident in the workplace. A guide for employers.* Document 2963C(01/05). Available at: http://www.wsib.on.ca/wsib/wsibsite.nsf/LookupFiles/DownloadableFileHandlingCriticalIncident/$File/Critical.pdf, retrieved January 14, 2009.

44. WorkSafeBC. *Critical incident response.* Available at: http://www.worksafebc.com/claims/srious_injury_fatal/wrk_05.asp, retrieved January 8, 2009.

45. Manitoba's Information CJOB|68. *WRHA Reports 32 "Critical Incidents."* A press release by the Winnipeg Regional Health Authority, Janury 6, 2009. Available at: http://www.cjob.com/News/Local/Story.aspx?ID=1050733, retrieved January 8, 2009.

46. College of Licensed Practical Nurses of Nova Scotia, College of Occupational Therapists of Nova Scotia, College of Physicians & Surgeons of Nova Scotia, College of Registered Nurses of Nova Scotia, Nova Scotia College of Pharmacists, Nova Scotia College of Physiotherapists. *Joint Position Statement on Patient Safety.* Available at: http://www.cpsns.ns.ca/2008-joint-patient-safety.pdf, retrieved January 27, 2009.

47. Woods DM, Johnson J, Holl JL, Mehra M, Thomas EJ, Ogata ES, Lannon C. Anatomy of a patient safety event: a pediatric safety taxonomy. *Qual Saf Health Care,* 2005;**14**:422–.

48. Chang A, Schyve PM, Croteau RJ, O'Leary D, Loeb JM. The JCAHO patient safety event taxonomy: a standardized terminology and classification schema for near misses and adverse events. *Int J Qual Health Care,* 2005;**17**(2):95–105.

49. Van der Schaaf TW. *New Miss Reporting in the Chemical Process Industry.* Eindhoven: Eindhoven University of Technology, 1992 (Thesis).

50. Battles JB, Kaplan HS, Van der Schaaf TW, Shea CE. The attributes of medical event-reporting systems. Experience with a prototype medical event-reporting system for transfusion medicine. *Arch Pathol & Lab Med,*1998;**122**(3):231–38.

51. Morimoto T, Gandhi TK, Seger AC, Hsieh TC, Bates DW. Adverse drug events and medication errors: detection and classification methods. *Qual Saf Health Care,* 2004;**13**:306–14.

52. Gandhi TK, Weingart SN, Borus J, Seger AC, Peterson J, Burdick E, et al. Adverse drug events in ambulatory care. *N Engl J Med,* 2003;**348**(16):1556–64.

53. The World Alliance for Patient Safety Drafting Group; Sherman H, Castro G, Fletcher M on behalf of the World Alliance for Patient Safety, Hatlie M, Hibbert P, Jakob R, Koss R, Lewalle P, Loeb J, Perninger T, Runciman W, Thomson R, van der Schaaf T, Virtanen M. Toward International Classification for Patient Safety; the conceptual framework. *Int J Qual Health Care,* 2009;**21**(1):2–8.

54. Thomson R, Lewalle P, Sherman H, Hibbert P, Runciman W, Castro G. Towards an International Classification for Patient Safety: A Delphi survey. *Int J Qual Health Care,* 2009;**21**(1):9–17.

55. Donaldson LJ. In terms of safety. *Int J Qual Health Care,* 2006;**18**(5):325–6.

56. Makeham MAB, Stromer S, Bridges-Webb C, Mira M, Saltman DC, Cooper C, Kidd MR. Patient safety events reported in general practice; a taxonomy. *Qual Saf Health Care,* 2008;**17**:53–7.

57. Anon. *Causal and Contributing Factors.* Available at: www.fire.ca.gov/fire_protection/downloads/esperanza_09_causal_contributing_factors,pdf, retrieved February 3, 2009.

Chapter 4

Describing Medical Error and Harm: Their Occurrence and Nature in Clinical and Community Settings

Executive Summary

Research in the medical error domain is polarized between not only system and individual visions (paradigms) but also the search for and detection of causes of error (that are already intuitively, by experience, or by etiological research proven causes) or the research that must demonstrate some feature, variable, or phenomenon as a "new" cause of error. Moreover, research often focuses on incidents, immediate harm, or other more distant outcomes consequent to the error itself as being synonymous to the medical error itself. Given the occurrence of error in unique, sporadic, and rare cases, even simple descriptions of medical error occurrence require several methodological approaches and considerations to arrive at the number of events, their outcomes, their severity, persons/time/ place characteristics, and what might be concluded, at least hypothetically, to be better demonstrated later.

Two types of descriptions of cases of medical error and harm are then considered: (1) unique, rare, or infrequent cases; and (2) series and larger sets of cases. Their occurrence is suitable for description based on epidemiological methods and techniques.

Error occurrence in unique or rare cases or case series may be described using the methodology of modern clinical case reporting. A complementary or alternative approach may be case-based qualitative research and narrative methods in quality improvement research, however their causal demonstration power may (and will) be.

Describing a medical error case series means creating a picture of the persons/time/place involved by specifying the state of ad hoc knowledge, research objectives, definition of the medical error under study itself and of the related variables, the study design, its target population, measurements and classifications to be used, and sources of data and then presenting a picture of events of medical error and its consequences, comparisons of findings, and conclusions about the picture created including causal hypotheses generated by simple observations supported at each step by relevant factual and methodological references from past experience and achievements.

Medical error occurrence studies often include a number of "soft" data, which are difficult to define, measure, and classify by means of typical methods in clinical epidemiology. An appropriate "hardening of soft data" should be applied wherever and whenever possible and necessary.

As in many other domains of medical research, the causal proof of causes of medical error and the causal proof of its consequences beyond most convincing descriptions remains an important challenge. Chapter 5 explores the search for causes in the medical error domain.

Thoughts to Think About

Harm watch, harm catch.

—**A proverb**

The errors which arise from the absence of facts are far more numerous and more durable than those which result from unsound reasoning respecting true data.

—**Charles Babbage, 1832**

Start out with the conviction that absolute truth is hard to reach in matters relating to our fellow creatures, healthy or diseased, that slips in observation are inevitable even with the best trained faculties, that errors in judgment must occur in the practice of an art which consists largely of balancing probabilities—start, I say, with this attitude in mind.... You will draw from your errors the very lessons which may enable you to avoid their repetition.

—William Osler, 1892

I keep six honest serving-men,
(They taught me all I knew);
Their names are What and Why and When and How and Where and Who.

—Rudyard Kipling, 1863–1933

The most erroneous stories are those we think we know best—and therefore never scrutinize or question.

—Stephen Jay Gould, 1997

Absence of evidence of harm is not evidence of absence of harm.

—Andrew Simms, 2003

It is not enough to collect and collate errors. Error trends need to be analysed particularly when the same mistake is encountered either repeatedly or across a number of facilities.

—John F. A. Murphy, 2009

More is missed by not looking than not knowing.

—Moshe Schein, 2003

A search for causes, prevention, and control must follow, we believe. Medical errors and medical harm are not exactly the same, and neither are their descriptions from a nonepidemiological or epidemiological standpoint.

Introductory Comments

Many practitioners and researchers do not have a high opinion of purely descriptive studies of health problems since they feel that these studies say nothing! The value of occurrence studies lies elsewhere. They offer the best possible picture of the magnitude of health problems (including medical errors), in various individuals, groups, times, and places. They also help set priorities as pertains to where to search for causes and where to intervene to minimize the problem if its causes are known. If the causes are unknown, these studies provide hypotheses regarding possible causes and in which direction etiological research should be oriented. The challenge of the medical problem and its understanding and control are all governed by the same rule. What kind of medical error problem exists in our healthcare environment, and how big is it? How is it spread? What are the next steps to manage it?

As in any other domain, the study and management of medical error and harm fall into three epidemiological categories:

1. The occurrence of the health problem is described: in whom, where, and when does it occur, and how does it evolve in time and space? *Descriptive epidemiology* is put to work.
2. By observation, sets of cases of interest or community groups are compared in a cross sectional or longitudinal manner to elucidate at least in part causes of health problems. *Analytical epidemiology* remains a cornerstone of cause–effect matters.
3. Clinical trials, community health programs, and other intentional and planned actions are used to strengthen our idea about cause–effect relationships and to demonstrate if preventive or curative interventions are an effective strategy to control a health problem—error and harm included. *Experimental epidemiology* or interventional epidemiology contributes to answering such important questions.

This chapter is about the first category of the aforementioned epidemiologies and how it might be usefully applied to and used in the medical error and harm domains.

Chapter 5 focuses on the second category (observation-based analytical), and Chapter 7 looks at the challenge of interventions in lathology.

Research, Knowledge Acquisition, and Intervention Strategies in the General Error Domain as Viewed by a Methodologically Minded Physician Epidemiologist

So far, our understanding of the error problem in general relies mainly on methodology provided by psychology, operational research, and ergonomics. However, we may think of medical error as a health phenomenon like disease, elements of healthcare, or behavior. As such, our understanding and controlling of the medical error challenge relies heavily on fundamental and clinical epidemiology methodology.

The general error methodology is based to a great degree on a priori knowledge of possible causal factors leading to an essentially inductive approach to problem solving. It is expected from epidemiology that the causal role of various factors must be established before any meaningful control can be considered. Consequently, inductive research may be a starting point for hypotheses generation (i.e., what causes this or that medical error) only. Deductive research remains a foundation from which we can build our knowledge of the etiology of error and its outcomes.

Table 4.1 compares and contrasts the "traditional" or prevailing approach that Perneger originally named the *causal attribution model*[1] and the epidemiological approach. It immediately shows that traditional and epidemiological strategies are complementary rather than contradictory.

Error in medicine, like any other health phenomenon such as disease, its causes, consequences, medical interventions, or other aspects of medical care, is worth studying, preventing, and controlling through epidemiological methods. It must be. In this context, the main challenges remain the definition of medical error itself and its variability in persons/time/place. Some errors are exceptionally rare and are rooted out quickly, like prescription errors of new drugs with narrow indications; others are frequent, like errors leading to needle-prick injuries in hospital practice. Some errors are known and defined as such a priori, and their study and interventions are tailored to the known problem. Other errors are labeled errors only after they occur; these are defined as a new unknown entity a posteriori (after the first event), and error and knowledge about them is rebuilt according to this kind of "operational" definition.

So far, we agree with Busse[2] that, in medicine, most of our experience and knowledge regarding human and other error is more about "what happened" and "how it happened" rather than "why it is happened." Analytical epidemiology should prove useful in better elucidating the yet unknown causes of medical error and harm.

Our epidemiological approach goes well beyond Dekker's view of the epidemiological model of accidents[3] as discussed in Chapter 2. It covers, among other things, three main domains, methods, and techniques of both field and clinical epidemiology: (1) *describing* what happens and *explaining* its causes and findings; (2) *putting into practice* and *evaluating* health interventions to prevent and control undesirable health problems; and (3) *implementing* and strengthening elements that improve human health and condition. In our case, this amounts to describing, explaining, and *controlling* medical error and harm.

In this spirit, we may see the classification of studies in the medical error and harm domain as illustrated by Figure 4.1.

Let us look first at the description of single, few, or rare cases.

Table 4.1   Distinctions between Prevailing and Epidemiological Strategies in General and Medical Error Domains

Subjects of Interest	Prevailing General Strategies	Epidemiology and Medical Error Model
Core disciplines	Psychology	Epidemiology
Direction of investigation	Operational research	Evidence-based approach
Causal model	Ergonomics	Uses of evidence
Entity under study	Information sciences	Deductive or inductive
Design of investigation	Inductive	A posteriori causal attribution
Assessment of causes	A priori causal attribution	Webs of causes
Key advantages	Linear sequence	Webs of consequences
	Error itself	Error as "disease"
	Outcomes of error: harm	Error as cause or consequence
	Severity and nature of outcomes	Occurrence studies
	Case studies	Case-control studies
	Single case qualitative research	Cohort studies
	Expert judgment	Hybrid studies
	Flexibility and insight	Clinical trials
	Sensitivity to context	Health program evaluation
	Exhaustive data and information	Fulfillment of causal criteria in epidemiology
	Attention to possible causal mechanisms plausibility	Generalizability

Table 4.1 Distinctions between Prevailing and Epidemiological Strategies in General and Medical Error Domains (*Continued*)

| | | Evaluation of joint effects from webs of causes |
| | | Quantification of strength and specificity of associations |

Source: Modified and expanded from Perneger, T. V., *Int. J. Qual. Health Saf.*, 17, 1, 2005.

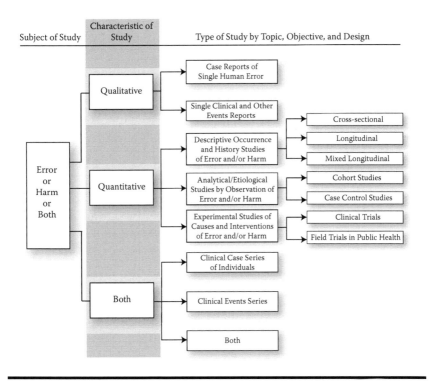

Figure 4.1 Type of study in the medical error and harm domain by topic, objective, and design.

Descriptions of Single Cases, Small Sets of Error Cases, and Harm Cases

Besides clinical case reporting in medicine, the methodology for describing single cases stems from other domains of qualitative research that also contribute to the methodological and terminological diversity we deal with not only in this chapter but also in descriptive studies of medical error and harm.

Choosing a Research or Intervention Subject

Ideally, any description in the error and harm domain or anywhere else in the health sciences should be well defined, comparable, and representative of the health problem, communities, and settings of healthcare, means, and practices. This requirement applies to all areas of inquiry and intervention.

Not everything is suitable for research and practice initiatives. Even in a purely descriptive context, clear questions must be formulated, data and information material must be solid, and analyses of relevant findings must produce valuable evidence. Such results must be usable for implementing corrective measures, must be generalizable, and must be evaluable. Figure 4.2 illustrates a cascade of necessary components and their worthiness in choosing what to work with. Such necessary components are particularly challenging in the domain of error and harm because they are "young" compared with historical and more traditional fields and application of epidemiology.

By bridging terminology and meanings in general and in medicine, we may look at the medical domain as a domain of "operations" (activities and events subject to error) run by "operators" (i.e., physicians, nurses, other professionals involved in clinical and community care using technologies and physical environments offered to them and their patients and executors of specific tasks).

Table 4.2 shows how research and interventions might differ according to the following:

- Subject of interest.
- Grounds for error.
- Operator adequacy.
- Core error.
- Action (error) outcomes.

Such variables may be treated as independent (potential causes of error) or dependent (error as consequence). In the study of the impact of error, the dependent or independent role in causal considerations is reversed.

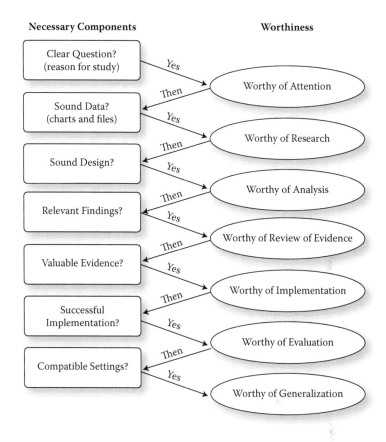

Figure 4.2 Necessary and required components and their worthiness in a cascade applicable to any descriptive study. (Adapted with modifications from Jenicek, M., *Foundations of Evidence-Based Medicine,* **Parthenon Publishing Group/CRC Press, Boca Raton, FL, 2003.)**

As we insist throughout this book, medical error and harm are entities that may or may not be related. Therefore, they must be treated separately and linked together later when such relationships really exist.

Descriptions of what happens in the area of medical error and harm fall into two major methodological categories:

1. Medical errors and harm are rare, exceptional, unique, or appear in small sets; in this case it is difficult to find other similar events. These situations are subject to clinical case reporting or qualitative research and case studies as seen in nonhealth domains. They may be reported and analyzed in a

Table 4.2 Error and Harm: Their Subjects, Grounds, and Focuses of Study

Subject of Study	Error Grounds	Purposes (Focuses) of the Error Study
Operation itself (terrain)	States breeding error	Condition of clinical, human, and physical environment
Operation target	Target conditions	Patient state and disposition besides diagnosis
Operator himself or herself	Inadequacy	Operator's general physical and mental disposition
	Inadequacy	Operator's knowledge and experience
Operator's error	Error itself	Operator's faults in reasoning (human error) argumentation, critical thinking
	Error itself	Operator's decision making
	Error itself	Operator's sensory-motor activity and action
Harm from error	Error outcomes	Outcomes for the • Patient (injury or death) • Physician (performance, health) • Health team (interaction, communication, actions) • Health system (man/machine) workings

"nonmedical" manner based on qualitative research methodology or may be examined in "clinical" case reporting, as is done today.

2. Error and harm events are more frequent and are subjects of an epidemiological inquiry based on experience with large sets of cases allowing descriptions of the occurrence of error or harm according to persons/time/place setting and characteristics.

Let us discuss them in this order while noting, however, that there is currently no clear limit between the numbers of events subject to qualitative research and the numbers of cases suitable for epidemiological description and analysis.

Epidemiologists make a distinction among three types of error and harm single-case or case series reports:

1. Cases are simply *counted*, enumerated, and presented with their relevant characteristics. Case series "without denominators" are grouped together.
2. The occurrence of cases (number of cases in the numerator of some rate) is *related to suitable "denominators,"* that is, numbers of individuals such as groups of patients receiving similar care, community groups, or other sets of individuals yielding "rates" of error or harm cases.
3. Case series of error or harm, with or without denominators, are *used for comparison* with some control groups in causal research in lathology.

These three types of reports are important in their own way, especially if their findings are not used beyond their essentially descriptive nature.

Reporting Unique, Infrequent, or Rare Cases beyond the Customary Methods of Clinical Practice: Case-Based Qualitative Research and Narrative Methods in the Area of Quality Improvement

What happened exactly when this medical error occurred? What were the elements of the clinical environment involved? How were the health system and working team functioning? What are the possible causes of failures? How did the physician reason and make decisions erroneously? What were the outcomes, consequences of reasoning, and decision-making failures as harm to the patient, system, and all participating actors? An in-depth inquiry into these matters and answers to such questions rely first on in-depth analyses of single-error case events before any quantitative (biostatistical and epidemiological) analysis and interpretation occur.

Within the framework of qualitative research, case studies are essential to our further understanding of medical error events. Let us briefly outline both of them.

Qualitative Research

Studying and reporting single cases of an important medical error should first be the subject of what we call today *qualitative research*. This type of research involves an in-depth study of an individual or situation, whereas *quantitative research* is rather a summary of sets of individual experiences. Qualitative research was simply and perhaps best defined by Strauss and Corbin[5] as "any kind of research that produces findings not arrived at by means of statistical procedures or other means of quantification.... It can refer to research about persons' lives, stories, behavior, but also about organizational, functional, social movements, or interactional relationships.... Some data may be quantified as with census data, but the analysis itself is a qualitative one."

The primary objective of quantitative research is to provide answers to questions that extend beyond a single observation. In qualitative research, unique observations are the focus of interest. They are described, studied, and analyzed in detail. They are not considered a part of a given universe before being linked to other observations. The objective of this kind of research is primarily to understand a case, a single observation itself. Pope and Mays[6] stress that "a given observation (X, case)" is the focus of qualitative research with classification as its purpose. By contrast, quantitative research enumerates how many Xs. The future of our understanding of medical error relies on a proper and balanced combination of qualitative and quantitative methods.[7,8]

Qualitative research in the medical error domain, like elsewhere, has as its objective an in-depth understanding of and the results that govern human behavior. In our case, the behaviors and results examined are those of health professionals and other partners in healthcare, even patients themselves in their physical and human environment settings and the events that cause medical errors.

Research questions can be approached by induction or by deduction. *Inductive research* is centered on observations that serve as a basis for hypotheses and answers, whereas *deductive research* raises questions, gathers observations for the problem, and confirms or rejects hypotheses free from or independent of the subject under study. A study of single cases of medical error is primarily of an inductive nature. Qualitative research focuses mainly on exploring and generating hypotheses.

While quantitative research or epidemiology generally tries to answer Rudyard Kipling-like questions such as when, where, how, in whom, and why, qualitative researchers contribute particularly to answer why or how questions by generating hypotheses about cause–effect relationships of interest. However, only "real" etiological studies like case control or cohort studies or experimental research (e.g., trial) will provide final answers to the why and how questions.

Qualitative studies are based on five approaches:[9]

1. *Biographical life histories,* or "telling the story of what happened to a single individual."
2. *Phenomenological studies,* which summarize lived experiences in several individuals.
3. *Grounded theory studies,* or "tell me as much as you can about this particular problem." Its purpose is to generate or discover a theory.

4. *Ethnography,* the process of describing and interpreting a cultural or social group or system.
5. *Case studies,* which explore one or more cases over time through detailed data.

Several valuable monographs now summarize experience and methodology in qualitative and mixed-method (combining qualitative and quantitative strategies) techniques in social and health sciences.[10-14] In 1995, the *British Medical Journal* introduced its readers to qualitative research in health sciences and in medicine, in particular, by means of an excellent reader-friendly collection of articles that later appeared in book form.[12,13]

Greenhalgh et al.[15] see narratives as the "science of imagination" joining the "science of concrete" of formal logico-scientific research. To present quality improvement research as a focused systematic inquiry that uses narrative methods to generate new knowledge, these authors propose the narrative interview, naturalistic story gathering, organizational case studies, and collective sense-making as four ways to enrich information provided by formal logico-scientific research.

Let us remember that qualitative research principles are not unknown to medicine. Medical history, psychiatric interviews, or the study of index cases (i.e., those that lead to a more expanded quantitative search) of disease outbreaks or of new phenomena (e.g., poisoning, infection) bear the characteristics of qualitative research. In practice, dealing with the medical error problem means that surveying, reporting, and compiling statistics about medical error are not enough if they are not followed by qualitative analyses of medical error first and quantitative research second.

Greenhalgh[16] suggests ways to read and understand "papers that go beyond numbers." Hence, qualitative research has gradually joined the mainstream of findings in health sciences. At the same time, qualitative research methodology, its relevance, and fields of application were presented to physicians in several introductory articles, series of papers, and books reviewed elsewhere.[17] The focus of those first initiatives was both on patients as cases and situations (services) as "cases." Randomized clinical trials, clinical interviews, patients' unmet expectations for care, or cases of injury can also be evaluated as cases in qualitative research.[17] Why should this not also apply for medical error?

Over the past decade or so, the interest in qualitative research in medicine has grown, with its methodology and applications suggestions becoming increasingly specific to various health professionals.[17-20]

Case Studies of Medical Error and Harm

A *case study* is an in-depth examination, analysis, and interpretation of a single instance or event. In this sense, a *case study of medical error* is the study of an *event* as a primary focus of interest. Patient, disease, services, and providers of medical care are all parts of such an event background and essential elements of analysis.

In general, a case study refers to the "collection and presentation of detailed information, looking intensely at an individual or small participant pool, drawing conclusions only about the participant or group and only in that specific context. Emphasis is placed on exploration and description, not on a cause-effect relationship." [21abridged] For Rothe,[22] "…the term 'case study' comes from the tradition of legal, medical, and psychological research, where it refers to a detailed analysis of an individual case, and explains the dynamics and pathology of a given disease, crime, or disorder. The assumption underlying case studies is that we can properly acquire knowledge of a phenomenon from intense exploration of a single example."

A case study can be qualitative (search for a meaning) or quantitative (some kind of measurement); for example, clinical examination as a case study is both qualitative (e.g., history taking, psychiatric evaluation) and quantitative (e.g., anthropometry, examination of vital functions).

In qualitative research, case studies are based on individuals or situations as cases.[23] This strategy explains case study uses in policy, political science, public health administration research, community psychology and sociology, organizational and management studies, or city and regional planning. Health institutions or agencies (e.g., health maintenance organizations, hospitals, neighborhood clinics) and their organization and functioning can also be studied as cases.[17]

Case studies may be[24]

- *Illustrative* (to accurately describe the problem).
- *Exploratory* (to identify methodology, questions, or other objectives of larger studies that will follow).
- *Cumulative* (to gather information from various sites and time periods as in error surveillance activities).
- *Critical instance* case studies (examining unique situations of special interest).

All of these types are relevant in the medical error domain. In health administration, medical care organization, and health services research, case studies are used in their expanded sense.[25] Incentives arising from de-insurance of in vitro fertilization, consequences of the introduction of public funding for midwifery, or patients' unmet expectations for care [26,27] can be quoted as examples.

Case studies are essentially of a qualitative nature.[28] Case studies begin and operate first with *anecdotal evidence*,[29] *narratives*,[30] or hearsay, all often based on insufficient evidence.[31] Case studies as research tools, like in any other research methods, now include the following:[6,15,32,33]

1. Formulation of a research question.
2. Selection of the cases.
3. Operational definitions of variables of interest.
4. Study design.
5. Definition of methods and techniques of data gathering and analysis: special attention is paid to the nature and procedure of interviews, interviewed subjects, transcription of interviews, and related ethical questions.
6. Logical linking of data to propositions.
7. Data analysis.
8. Criteria for interpretation.
9. Interpretation of findings.
10. Reporting of findings.

A qualitative case study of a medical error must respect the same rules. Within such general rules, the following mainly epidemiological challenges must be specified:

■ The inductive or deductive nature of the inquiry.
■ Whether interviews will be based on open questions, direct questions, or both.
■ What data will be considered hard (i.e., as defined beforehand, measurable, reproducible), what data will be considered *soft* (e.g., fuzzy definitions of sorrow, pain, indigestion, and the like), and what will be done to harden (i.e., make more scientifically usable) soft data.[34]
■ How to ensure interviews are as objective as possible and reflecting reality beyond a single inquirer and inquired.

To improve the quality of findings, qualitative researchers use, among others, repeated inquiries by the same person and different persons, based on different methods and techniques of data and information gathering and analyses, called *triangulation*. This term, meaning "looking at one thing from different angles," is derived from the field of navigation and military strategy.

So far, authors of medical errors reports do not always specify if their reports represent an exercise in qualitative research and case study in the aforementioned senses. Future experience in this direction will be another improvement in understanding, preventing, and controlling human error in medicine.

Two Examples of Qualitative Research in Medicine and in the Domain of Medical Error

Two examples of qualitative research illustrate recent initiatives in the health domain:

1. In general healthcare, the *Engaging Physicians, Benefiting Patients* study[35] for the U.K. Department of Work and Pensions has as its objective to evaluate the reported impact of the pilot communication strategy and to contribute to the knowledge about general practitioner attitudes toward their role of provision of medical evidence in general. It also includes considerations of broader aspects of culture, attitudes, and patient–doctor relationships. The study is based on interviews with both physicians and practice managers in a pilot area.

2. In the domain of medical errors in primary care, closer to our subject, Kuzel et al.[36] pay attention to the development of patient-focused typologies of medical errors and inquiries to understand the seriousness of errors from a patient perspective; to compare such patient perceptions with those of primary care physicians; to provide a basis for further error research in terms of typology, causes, and occurrence (epidemiology) of adverse events; and to construct models and simulations for research and practice to correct and improve medical error in primary care.

Such initiatives should prove even more valuable in relation to other future projects.

Reporting Single Cases of Error and Harm the "Medical" Way

There are two ways to report a medical error:

1. The first may be a narrative report simply telling what happened and only later on analyzing the situation and proposing remedies to it. This is the strategy, for example, chosen by the U.S. Agency for Health Care and Quality (AHRQ). This agency publishes an excellent and meritorious *Case Archive* series[37] online covering single-case medical error and harm happenings mostly in a clinical setting. Its case reports of harm are generally narrative and descriptive, covering both error and harm and their management as well as their discussion and analyses by lathology experts as third (reference) parties.

2. We may (and should) also consider medical error as the subject of a more structured clinical case report as we understand it today.[38,39]

In this second case, the architecture of the clinical error and harm case report should, like anywhere else, include the following:[38,39]

■ A *summary* (reasons for the report, background of the problem, most relevant conclusions).

- An *introduction* (definition and context of the error and events related to it, objectives and justification of reporting this error case).
- *Presentation of the error case* (what happened, including important patient, physician, clinical environment, and medical care factors relevant for understanding the error problem in this case). Describe the persons involved and their activities; system components and functioning; and physical, biological, or social environment and setting relevant and necessary for understanding the error happening and its causes.
- *Discussion and conclusions* about how to understand the problem in this specific and also more general context and what remedies should be considered for this error occurrence. Specify where your error of interest belongs in the current taxonomy you prefer and need: human error; system error; accident; incident or critical incident; lapse, mistake, or slip; active or latent; cognitive or decision affecting; skill-, rule-, or knowledge-based.
- *References* that include past experience and methodological aspects related to this error case, relevant also for further work in similar situations.

If the medical error report is the result of a qualitative study, its architecture does not differ from that of a quantitative or mixed-method study: title, authors and site of the study, structured abstract, introduction, aims of the study, review of the literature, sample, data collection methods, data analysis methods, findings, discussion, conclusion, references. Uniform requirements for manuscripts submitted to biomedical journals also follow this structure.

Other additional key criteria for qualitative research and for reporting it well have been suggested as follows:[40]

- Presenting one's own perspective, and reflecting on subjectivity and bias making coherent connection between theory and method.
- Focusing on meaning.
- Accounting for, and being sensitive to, context.
- Open-ended stance on data collection and analysis.
- Collection of and in-depth engagement with "rich" data.
- Balancing description and interpretation of data.
- Offering transparent analysis (e.g., grounded in example).
- Offering plausible, credible, and meaningful (to reader and others) analysis.
- Offering a sense of what is distinct within the account of what is shared.
- Drawing out relevant conclusions.

Other details regarding qualitative data and findings reporting and requirement for organization, coverage, knowledge and understanding, logical development, evaluation, originality, writing style, initiative and effort, and research skills may be found elsewhere.[40-43]

Reporting Case Series of Error and Harm

In this case, a medical error case series report should include all necessary elements and characteristics of an occurrence study in epidemiological sense, frequency of cases, persons/time/space characteristics, ensuing hypotheses for the understanding of the problem, and directions for further study and practice (without exaggeration in cause–effect considerations).[44]

Whenever possible, not only frequency of error events but also their denominators should be reported. Denominators of their rates may be numbers of related healthcare acts, number of patients, or other frequencies depending on questions asked. Incidence rates or incidence densities in relation to various persons/time/space characteristics are necessary for further work on the error problem.

Epidemiologists teasingly be called "human beings in an eternal search and quest for correct denominators" (i.e., for rates of disease or death they describe and analyze). They continue to search in the domain of lathology. For example, in the field of error, error rates having hospital admissions as denominators are useful for basic descriptive purposes and to define the magnitude of the problem in a particular setting. For additional analytical purposes and causal search, denominators as frequencies of the same acts (e.g., laparoscopic surgeries, medications for a specific condition) that also include all the same events when "nothing happened" remain underused from an epidemiological perspective. Also, we are keeping in mind that case series obtained now may be used later in some type of causal evaluation by adding to the case series some kind of historical or concurrent control groups.

This concept of an essentially "descriptive" study of error in medicine will necessarily include at least 12 of the following common characteristics of a descriptive study of disease cases or other clinical or community medicine events:[44]

1. Current and desired acquisition of *knowledge* about the error problem to be described (with referenced past experience).
2. Research *question and objectives* of this essentially descriptive study.
3. *Definition of the medical error under study* (in the best operational terms possible).
4. *Definition of variables possibly related to it* as causes, risk factors or markers, or its consequences (outcomes).
5. Identification of the *study design* (e.g., cross sectional, longitudinal).
6. Specifying the *target population* of the study (e.g., patients, health professionals, setting of care, community).

7. Clinimetric information like *measurement* and *classification* of error and clinical and other events related to it.
8. Identification of *sources of data.*
9. Presentation of the *picture of error events* (*epidemiological description,* which includes all numerators and denominators needed for the establishment of the magnitude of the problem and its distribution and further etiological research).
10. *Comparison of findings* with other findings or experience (if available).
11. *Conclusions about the observed error series* (is the problem still sporadic or already epidemic, endemic, or pandemic, with hypotheses drawn to interpret the problem, its cases, and possible remedies). Conclusions are those that can be derived from this study only in relation to the past experience. Generating hypotheses from a descriptive study is not a substitute for an analytical etiological study!
12. *Factual and methodological references* from past experience and literature.

In this spirit, a descriptive study says something and represents a precious experience enriching our knowledge of the medical error domain.

When health professionals decide to study and report medical error and harm by way of qualitative research, they quickly realize that reporting in medical journals has its own requirements, identical in many respects to those of quantitative research. Rowan and Huston[45] presented in the *Canadian Medical Association Journal* (*CMAJ*) a checklist for authors and peer reviewers of qualitative studies. Abridged here, it includes, within a more traditional structure, typical requirements for qualitative and quantitative study reporting and several specific elements inherent to the rapidly developing qualitative research methodology:

1. In the *Introduction* section, the relevance of the research topic is emphasized, supported by an appropriate literature review and leading to a clear and structured research question. The ethical approval of the research is documented.
2. In the *Methods* section, a qualitative approach appropriate for the question at hand is justified. The setting description facilitates the understanding of a rational sampling method including sample size development and calculation. The collection of information and the analysis that follow focus on the understanding, dependability, clarity, appropriateness, and transferability of findings.
3. The *Findings* section offers a clear description of what was seen, heard, and recorded in a confidential manner. Interpretations flow logically from the insightful analysis of findings and their answer to, ultimately, the research question at the origin of the study.

4. The *Discussion* section does not omit implications and alternative interpretations of results, strengths and weaknesses of the study, and directions for further inquiry.

To this list, let us add that the *References* should cover factual information already available in other studies on the same or similar topic, the methodology of qualitative research used, and its links and how they complement and complete the quantitative information and methodology available in the domain of interest.

Back to Epidemiology: What Happens Now? Occurrence Studies, Descriptive Epidemiology, Magnitude, and Distribution ("in Whom, Where, and When") of the Error and Harm Problem

Descriptive epidemiology or the study of error occurrence (frequency and distribution) remains limited for two major reasons. First, errors with their often dramatic consequences are fortunately rare, if not exceptional; consequently, statistically meaningful data are equally scarce. Second, definitions of error itself still abound; they are not all of an operational nature, having being made for different purposes of study and control. Ensuing multiple taxonomies of error facilitate neither description nor analysis of the error problem.

International, national, state, or local reporting systems and inquiries now cover both hospital[46] and extrahospital[47] practices. These studies are mostly descriptive.[48] A descriptive study of an error problem in medicine must, like any other descriptive study (or study of disease occurrence), provide answers to several questions:

- How many (errors occurred)?
- To what outcomes do they lead?
- How serious (in terms of their outcomes)?
- Where, in whom, and when did they occur?

Most studies of medical error currently available are occurrence studies in the epidemiological sense. They focus on what happens in hospital care and in primary care.

A Short Epidemiological Reminder

There are perhaps two important points to stress, especially for any reader less familiar with epidemiology: (1) distinctions between an incident and incidence;

and (2) how incidence itself is presented in the medical error and harm domain. *Hazard* is another term with two meanings.

Incident and Incidence

However similar they may sound, *incident* and *incidence* are two different things. Risk and hazard as probabilities of a health event are different, too.

As already mentioned in Chapters 2 and 3, an ***incident*** is a process, event, practice, or outcome noteworthy for its potential to create hazard or harm.

The broadest lexical definition of ***incidence*** is that of "the range of occurrence or influence of something, especially of something unwanted"[49] or "degree, extent or frequency of occurrence, amount."[50] In simple terms, incidence means in epidemiology[51] the number of new events (e.g., disease cases) occurring during a given period of time. *Incidence rate* is such a number related as a numerator to some denominator represented by a population at risk, that is, these individual who may develop the problem or event during that period. Numerators are given a size suitable for comparisons, like per 1,000 or 100,000. If individuals in denominators are weighted by each and others' periods of time of exposure to some cause or by the duration of their follow-up, an *incidence density* is obtained.[4] This way, the expression of incidence in the medical error and harm domain may yield persons versus time of follow-up; persons versus time or number of work shifts of a health professional; persons versus days of hospitalization; persons versus number of medical acts potentially leading to an error or harm. It all depends on what is studied. A *proportional rate* may still occur when referring to "pie chart" statistics. Saying that 20% of all medical errors are of a surgical nature is a proportion of the total of some events and is hence a "pie statistic." (Epidemiologically speaking, there are neither denominators nor rates here.) Saying that the incidence rate of harm caused by medical error is 20% means 20 cases of harm occurred during a defined period in 100 subjects exposed (at risk) to such an error event over the period of interest.

Currently, as Battles and Stevens[52] suggest, the reader of studies of medical error and harm may find studies predominantly reporting numerators of epidemiological rates only[53,54] or reports mainly presenting proportional rates of error or harm.[55,56] In addition to that, distinctions between error and harm like in Hickner et al.'s study of family medicine practices[57] are becoming a necessary virtue. Shojania[58] reminds us that denominators are sorely missing in current studies of incident reporting.

An incident report like the report by Hoffman et al.[59] and others opens the door for a more complete study of incidence rates, customary in classical epidemiology. Dean et al.[60] show that prescribing errors in hospital inpatients provides an idea about the incidence of such events; distinctions between various expressions of incidence in epidemiological terms related to comparable periods of time, number of risk-generating acts, and other "densities" will certainly follow and should be encouraged.

Risk and Hazard

Risk in epidemiology is the expression of the probability of developing a new health problem over a certain period of time, either in an individual or in a group of individuals. A *hazard*, often synonymous with risk, is a factor or exposure that may adversely affect health.[51] Risk is more often used as an expression of the probability to develop a health problem in individuals who still do not have it, whereas hazard is often used as a term for a similar probability in the domain of prognosis (e.g., of an adverse development, complications) in subjects who already have the disease.[4]

Confusion between the two terms still occurs. In addition to such confusion, the term *hazard in lathology* (as already quoted in Chapter 2) is defined as the error capable of causing harm or a potential source of harm. It is a more general term standing for both danger (from illness) and risk (from interventions). Given the still persisting difference in definition of many such terms in epidemiology and lathology, any study of medical error and harm will benefit from specifying the meaning of such terms used in the study in question.

Error and Harm Reporting in Hospital Care

Pooling results from the Harvard Medical Practice Study[61,62] and the Utah-Colorado Medical Practice Study,[63] Studdert et al.[64] estimate that from their 17,192 observations of adverse events 7,715 (44.9%) are operative adverse events, 3.325 (19.3%) are drug related, 2,315 (13.5%) are medical procedure related, 1,181 (6.9%) are diagnosis related, and the rest are connected to other mixed disease or intervention backgrounds.

In the Utah-Colorado study, about 52% of adverse events are considered preventable. Baker et al.[65] found an overall incidence rate of adverse effects in Canadian hospitals in 2000 of 7.5%, suggesting that out of almost 2.5 million annual admissions, about 185,000 are associated with adverse effects and close to 70,000 of these are potentially preventable. The Canadian Institute for Health Information (CIHI) considers that medical error affects nearly 25% of Canadians, unevenly distributed from one Canadian province to another.[66] De Vries et al.[55] estimate on the basis of their systematic review of the international literature covering adverse events in hospitals that the median overall incidence of in-hospital adverse events was 9.2% with a median percentage of preventability of 43.5%; 7.4% of events were lethal. A total of 1 of 10 patients admitted suffers some kind of adverse effect of medical care. Surgery and medicine (in that decreasing order) are the major producers of adverse effects among medical specialties and fields of activity. Four-fifths of adverse effects occur in hospitals and the rest in various out-of-hospital fields of care.

Methodologically speaking, most reported rates are proportional rates rather than true incidence rates; even prevalence rates are sometimes mentioned (if the study period is hard to determine) like in Otero et al.'s Argentinean study of medication errors in pediatric inpatients.[67] More exact true incidence rates are less known, and incidence densities lesser still.

Additional information on medical error occurrence is covered by several Web sites, some of which were mentioned in 2001 by Sinclair.[68]

Error and Harm Reporting in Primary Care

An international group of primary care researchers formed the Linnaeus Collaboration with the goal of developing methods of primary care lathology (study of error in community-based primary care practice).[68,69] In Australia, Canada, England, The Netherlands, New Zealand, and United States, reported errors occurred between June and December 2001. In summary, Linnaeus findings show the following in Canada and other countries, respectively:

- 29% (Canada) and 39% (other countries) of errors are office administrative processes–related.
- 19% and 16% are due to external (paraclinical) diagnostic and exploratory procedures.
- 26% and 24% are connected to medications.
- 9% and 15% are due to communication failures between physicians and their patients.
- 5% and 3% are related to accounting and office management.
- 13% and 22% are caused by clinical knowledge failure which, in this study, meant not following the standard practice.

The most affected age group was 50–79-year-old patients, 64% and 57% of which were women. It was found that 40% and 41% of patients suffered from complex health problems and that 63% and 60% had problems of chronic nature. Among factors contributing to errors, in decreasing order, were process-, provider-, environment-, and patient-related factors.

Sandars and Esmail's[69] and Rosser et al.'s[70] assessment of studies of primary care in the United Kingdom, the United States, Netherlands, Australia, and Sweden show that medical error occurs between 5 and 80 times per 100,000 consultations, mainly related to diagnostic and therapeutic processes. Occurrence data gathering is still not uniform and varies from self-reports by health professionals to medico-legal databases and other sources.

It is not surprising that, in the relatively young field of lathology, the search for causes remains even more challenging than occurrence research. Let us now discuss etiological research in the domain of human error in medicine.

Guidelines for Describing and Reporting Medical Error and Harm Occurrence

Despite the diversity of definitions, classifications, causes, consequences, and settings of human error in medicine, we must see and understand this phenomenon in its spread, causes, and control as any other health problem of interest, be it a disease or any other health condition and its management. In this spirit, we must understand reporting of a single-error case or series of error cases.

From 36 reasons (and counting) to present a clinical case report reviewed elsewhere,[39] the following situations and reasons (as originally formulated) are particularly worthy of reporting in a medical error case:

- Clinical failures and unexpected disasters like mistakes in diagnosis or treatment and their consequences.
- Diagnostic and therapeutic "accidents" (causes, consequences, remedies).
- Unusual setting of medical care (e.g., emergency or field conditions) as potential breeding grounds for error.
- Emergency, often heuristic management of a case with or without necessary evidence, data, and information (or any other high risk of error situations and environments).
- Complications of surgery (adverse effects, rejections, other failures, intolerance of procedures).
- Presentation of "Black Swan" evidence: all swans are white; this unique black swan (a disease, its causes, and interventions) proves otherwise. In other words, it may be another potential error index source.

Any error and harm case presentation must be based on valid considerations and worthy, necessary components of such reports as summarized in Figure 4.2.

If a chronological picture of the case—its "natural history," clinically speaking—is of interest, an error and harm case report or description of similar events may be considered in the sequence illustrated by Figure 4.3.

Steps in such a sequence require, besides the general rules of clinical case reporting already outlined, the addition of elements specific to the error and

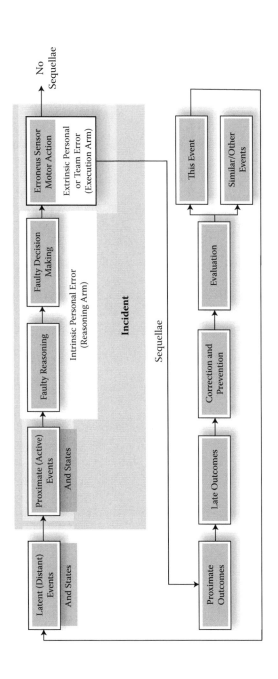

Figure 4.3   Error events—the general history.

harm domain, such as developments from incident to critical incidents or a subset of stages of error development followed by a subset of harm development, completed by steps in its control and overall evaluation of the entire evolution of the error and harm case.

The eight considerations and steps through which an incident evolves into a critical incident should be analyzed:

1. The *general setting of an error case (or cases)* is defined as an event, persons (patients and health professionals) involved, time, clinical physical environment, technological and pharmacological armamentarium in use, functioning of the services, and care.

2. The *potential or real latent causes* (we do not always know with certainty, but past experience, plausibility, and acquired knowledge will prevail), either of human or physical (environmental) origin.

3. *Proximate or active events, causes, and states* are mainly of human origin. Both latent and proximate events are evaluated as potential causes or consequences. Their taxonomies may differ. For example, Rasmussen's classification[71] includes causes of human malfunction (external and operator's); mechanisms of human malfunction (situation and performance shaping); and internal human and external modes of malfunction, both related to the personal task.

4. *Medical and healthcare error(s)* occur in a three-step process in which faulty reasoning is followed by faulty decision making and an erroneous sensory-motor execution of the task. They may be of human or system origin.

5. If an incident becomes critical, both unwanted *proximate* (injury) and *late* (frequency, spectrum of extent, gradient of severity) *outcomes* are retained for *decisions* about their *correction and prevention* and about the entire event.

6. If known and executed, corrective and preventive *measures proposed and implemented are evaluated* for their effectiveness, efficacy, and efficiency. Control of outcomes and their causes is applied to the best possibilities and means available.

7. An answer is sought to the question of whether findings from this experience might be *expanded, applied, and generalized* to other similar (or not) events and their settings.

8. An *additional reevaluation of the original setting* of occurring error and harm and their natural history may be of interest in such a circular process.

Steps 2–4 outline the medical error and incident story; steps 5–7 are about both medical harm and error and their control.

Conclusion

Reporting cases of medical error must then fulfill several requirements:

- Single cases and events will be reported in their structure and content like any other clinical case report.
- Given frequent uses of soft data, their best definition and handling are required.
- In most cases, fortunately, infrequent happenings of medical error do not allow a full deployment of prevalent epidemiological and biostatistical methods and techniques of disease (error) inquiry.
- Further uses, applications, development, and adaptation to the error problem of the qualitative research and case-study methodology will enrich our understanding of the medical error problem and will identify needs for further etiological and intervention research.
- Conclusions and recommendations stemming from single case or case series of medical error will not be exaggerated; cause–effect relationship proofs are reserved for other types of inquiry.
- On the other hand, creative and realistic hypotheses generation with indications and directions on what to do next to prove or refute them as well as implications for practice will not be omitted.
- The reasons for what is impossible and cannot be done realistically will be offered in balance with reasons and recommendations related to what is possible and what should be done.
- In contrast, "nonepidemiological" studies must not be swept away as something that "brings nothing" to the solution of the medical error problem. They may open the door to more advanced research, a desired direction, and may provide rich ideas worthy of pursuing in the not-so-distant future.

Even a simple description of a case of medical error and harm as discussed in this chapter as well as the analyses and corrective and preventive interventions described in the chapters that follow must be based on several critical thinking criteria—things to do and not to do,[72] as summarized in Table 4.3. Both the structure and the content of error and harm descriptions must be critically analyzed and presented in the framework of modern informal logic and critical thinking.

Epidemiologically speaking, the greatest challenge of descriptive or occurrence studies of medical error and harm remains their *representativity* (i.e., completeness or good sampling) and *comparability* with other observations and findings. Barriers to nationwide reporting of error and harm, mandatory or not, remain.[73] Is it possible to go beyond a pure description of the error problem?

Table 4.3 What to Do and Not to Do in Producing and Presenting an Error and Harm Inquiry in Light of Critical Thinking

Things to Do	Things Not to Do
Accurately interpret, for example, evidence, statements, graphics, and questions.	Offer biased interpretations of evidence, statements, graphics, questions, information, or the points of view of others.
Identify the salient arguments (reasons and claims) for and against.	Fail to identify or hastily dismiss strong and relevant arguments.
Thoughtfully analyze, evaluate, and present obvious and major alternative points of view.	Ignore or superficially evaluate obvious alternative points of view.
Draw warranted, judicious, and no-fallacious conclusions.	Not justify results or procedures or explain reasons.
Justify key results and procedures; explain assumptions and reasons.	Regardless of the evidence or reasons, maintain or defend views based on self-interest and perceptions.
Fair-mindedly follow where evidence and reasons lead.	Exhibit close-mindedness or hostility to reason.

Source: Adapted and reworked from Facione, P. A., in *Critical Thinking and Clinical Reasoning in the Health Sciences: An International Multidisciplinary Teaching Anthology*, Milbrae: California Academic Press, LLC, 2008.[72]

Deriving causal associations from single or infrequent event observation remains a subject of debate.

In this chapter, we have covered only basic definitions, classifications, and ways to describe medical error events. Let us now see in the next chapter how to analyze and explain medical error and harm events.

Murphy reminds us correctly that "'the error industry' must move to the next level. It is not enough to collect and collate errors. Error trends need to be analysed particularly when the same mistake is encountered either repeatedly or across a number of facilities. Recommendations for preventing future errors need to be proposed and appropriately funded where necessary."[74] Analysis means, in this context, an identification of causes by observational or experimental means wherever and whenever feasible. The next chapter calls for some epidemiological solutions. Is rational prevention and control of medical error possible without the best possible demonstration of cause–effect relationships? We believe not.

References

1. Perneger TV. Investigating safety incidents: more epidemiology please. *Int J Qual Health Saf,* 2005;**17**(1):1–2.
2. Busse DK. *Cognitive Error Analysis in Accident and Incident Investigation in Safety-Critical Domains.* PhD Thesis submitted at the University of Glasgow Department of Computing Science. Glasgow: University of Glasgow, September 2002. Available at: http://www.dcs.gla.ac.uk/~juhnso/papers/Phd_DBusse.pdf, retrieved October 25, 2008.
3. Dekker SWA. *The Field Guide to Understanding Human Error.* Aldershot, UK: Ashgate Publishing Limited and Ashgate Publishing Company, 2006.
4. Jenicek M. *Foundations of Evidence-Based Medicine.* Boca Raton, FL: Parthenon Publishing Group/CRC Press, 2003.
5. Strauss A, Corbin J. *Basics of Qualitative Research. Grounded Theory Procedures and Techniques.* Newbury Park, CA: Sage Publications, 1990.
6. Pope C, Mays N. Reaching the parts others cannot reach: an introduction to qualitative methods in health and health services research. *BMJ,*1995;**311**:42–5.
7. Goering PN, Streiner DL. Reconcilable differences: the marriage of qualitative and quantitative methods. *Can J Psychiatry,* 1996;**41**:491–7.
8. *Handbook of Mixed Methods in Social & Behavioral Research.* Edited by A Tashakkori, C Teddle. Thousand Oaks, CA: Sage Publications, Inc., 2003.
9. Creswell JW. *Research Design. Qualitative & Quantitative Approaches.* Thousand Oaks, CA: Sage Publications, 1994.
10. Corbin J, Strauss A. *Basics of Qualitative Research.* 3rd Edition. Thousand Oaks, CA: Sage Publications, Inc. 2008.
11. Creswell JW. *Qualitative Inquiry & Research Design. Choosing Among Five Approaches.* Thousand Oaks, CA: Sage Publications, Inc., 2007.
12. *Qualitative Research in Health Care.* Edited by N Mays, C Pope. London: *BMJ* Publishing Group, 1996.
13. *Qualitative Research in Health Care.* 3rd Edition. Edited by C Pope, N. Mays. Malden, MA: Blackwell Publishing and BMJ Books Publishing Group, Ltd, 2006.
14. Yin RK. *Case Study Research. Design and Methods.* 3rd Edition. Thousand Oaks, CA: Sage Publications, Inc., 2003.
15. Greenhalgh T, Russell J, Swingelhurst D. Narrative methods in quality improvement research. *Qual Saf Health Care,* 2005;**14**:443–9.
16. Greenhalgh T. *How to Read a Paper. The Basics of Evidence Based Medicine.* 2nd Edition. London: *BMJ* Publishing Group, 2001. (Pp. 166–178: Papers that go beyond numbers (qualitative research).)
17. Qualitative research. Section 2.3.4 (pp. 28–31) in: Jenicek M. *Foundations of Evidence-Based Medicine.* Boca Raton, FL: Parthenon Publishing Group (CRC Press), 2003.

18. *Qualitative Methods in Health Services Research.* Devers KJ, Sofaer S, Rundall TG (Guest Editors). *Health Services Res,* 1999;**34**:1083–1263.

19. Gantley M, Harding G, Kumar S, Tissier J. *An Introduction to Qualitative Methods for Health Professionals.* London: Royal College of General Practitioners, 1999.

20. Rothe JP. *Undertaking Qualitative Research. Concepts and Cases in Injury, Health and Social Life.* Edmonton: University of Alberta Press, 2000.

21. Colorado State University. *Writing Guides. Case Studies. Case Study: Introduction and Definition.* Available at: http://writing.colostate.edu/guides/research/casestudy/pop2a.cfm, 1 page, retrieved January 9, 2009.

22. Rothe PJ. *Qualitative Research. A Practical Guide.* Heidelberg, ON: RCI/PDE Publications, 1993.

23. *What Is a Case? Exploring the Foundation of Social Inquiry.* Edited by CC Ragin, HS Becker. Cambridge, UK: Cambridge University Press, 1992.

24. Colorado Sate University. *Writing Guides. Case Studies.* Available at: http://writing.colostate.edu/guides/research/casestudy/com2b1.cfm, retrieved January 9, 2009.

25. Keen J, Packwood T. Case study evaluation. *BMJ,* 1995;**311**:444–6.

26. Kravitz RL, Callahan EJ, Paterniti D, Antonius D, Dunham M, Lewis CE. Prevalence and sources of patients' unmet expectations for care. *Ann Int Med,*1996;**125**:730–7.

27. Inui TS. The virtue of qualitative *and* quantitative research. *Ann Int Med,* 1996;**125**:770–1.

28. Wikipedia, the Free Encyclopedia. *Case study.* Available at: http://en.wikipedia.org/wiki/Case_study, retrieved December 8, 2008.

29. Aronson JK. Unity from diversity: the evidential use of anecdotal reports of adverse drug reactions and interactions. *J Eval Clin Pract,* 2005;**11**(2):195–208.

30. *Narrative Based Medicine. Dialogue and discourse in clinical practice.* Edited by T Greenhalgh, B Hurwitz. London: BMJ Books, 1998.

31. Wikipedia, the Free Encyclopedia. *Anecdotal evidence.* Available at: http://en.wikipedia.org/wiki/Anecdotal_evidence, retrieved November 8, 2008.

32. Colorado Sate University. *Writing Guides. Case Studies. Case Studies.* Available at: http://writing.colostate.edu/guides/research/casestudy/com2b3.cfm, retrieved January 9, 2009.

33. Qualitative Research Inquiry for Psychologists (2003). *Writing a Qualitative Research Report. General Guidelines for Writing Qualitative Reports.* Available at: www.hopelive.hope.ac.uk/psychology/leveli/rsw/Qualitative_Report.doc, retrieved January 12, 2009.

34. From clinical observation to diagnosis: Clinimetrics. Section 6.1, pp. 109–113 in: Jenicek M. *Foundations of Evidence-Based Medicine.* Boca Raton, FL: Parthenon Publishing Group (CRC Press), 2003.

35. Hiscock J, Hodgson P, Peters S, Westlake D, Gabbay M. *Engaging physicians, benefiting patients: a qualitative study.* Department for Work and Pensions. Research Report No 256. London: Department for Work and Pensions, 2005. Available at: http://www.dwp.gov.uk/asd/asd5/rports2005/rrep256.pdf, retrieved January 12, 2009.

36. Kuzel AJ, Woolf SH, Engel JD, Gilchrist VJ, Frankel RM, LaVeist TA, Vincent C. Making the case for a qualitative study of medical errors in primary care. *Qual Heath Res,* 2003;**13**(6):743–780.

37. Agency for Healthcare Research and Quality (AHRQ). *Case Archive.* Available at: http://www.webmm.ahrq.gov/caseArchive.aspx, retrieved July 22, 2009.

38. Jenicek M. *Clinical Case Reporting in Evidence-Based Medicine.* 2nd Edition. London: Arnold and Oxford University Press, 2001.

39. Jenicek M. Clinical case reports and case series research in evaluating surgery. Part II. The content and form: Uses of single clinical case reports and case series in surgical specialties. *Med Sci Monit,* 2008;**14**(10):RA149–RA162.

40. Anon. *Features of a good qualitative project.* Available at: http://www.psy.dmu.ac.uk/michael/qual_good_project.htm, retrieved January 12, 2009.

41. Chenail RJ. Presenting qualitative data. *Qual Rep,*1995,**2**(3), Dec 1995, Available at: http://www.nova.edu/ssss/QR/QR2-3/presenting.html, retrieved January 12, 2009.

42. Anon. *Advice on writing up a qualitative study.* Available at: http://www.psy.dmu.ac.uk/michael/qual_writing.htm, retrieved January 12, 2009.

43. Burnard P. Writing a qualitative research report. *Acc Emerg Nursing,* 2004;**12**:176–181. Reprinted with permission from *Nurse Education Today,* 2004, **24**;(3):174–179.

44. Chapter 7. Describing what happens. Clinical case reports, case series, occurrence studies. Pp. 147–181 in: Jenicek M. *Foundations of Evidence-Based Medicine.* Boca Raton, London: Parthenon Publishing Group (CRC Press), 2003.

45. Rowan M, Huston P. Qualitative research articles: information for authors and peer reviewers. *CMAJ,* 1997;**157**(10):1442–1446.

46. Beasley JW, Hamilton Escoto K, Karsh B-T. Design elements for primary care medical error reporting. *Wisconsin Med J,* 2004;**103**(1):56–59.

47. Makeham MAB, Dovey SM, County M, Kidd MR. An international taxonomy for errors in general practice: a pilot study. *MJA,* 2002;**177**:68–72.

48. Weingart S. Epidemiology of medical error. *BMJ,* 2000;**320**:774–777.

49. *Random House Webster's Unabridged Dictionary.* Electronic edition V 3.0 fro 32 bit Windows. Antwerpen: Lernout and Houspie, 1999.

50. *Collins English Dictionary. Complete and Unabridged.* 6th Edition. Glasgow: Harper Collins Publishers, 2003.

51. *A Dictionary of Epidemiology.* 5th Edition. Edited for the International Epidemiological Association by M Porta, S Greenland and JM Last Associate Editors. Oxford: Oxford University Press, 2008.

52. Battles JB, Stevens DP. Adverse event reporting systems and safer healthcare. *Qual Saf Health Care,* 2009;**18**(1):2.

53. Braithwaite RS, DeVita MA, Mohindhara R, Simmons RL, Stuart S, Foraida M, and members of the Medical Emergency Response Improvement Team (MET) responses to detect medical errors. Use of medical emergency team (MET) responses to detect medical errors. *Qual Saf Health Care,* 2004;**13**:255–259.

54. Griffin FA, Classen DC. Detection of adverse events in surgical patients using the Trigger Tool approach. *Qual Saf Health Care,* 2008;**17**:253–258.

55. de Vries EN, Ramrattan MA, Smorenburg SM, Gouma DJ, Boermeester MA. The incidence and nature of in-hospital adverse events: a systematic review. *Qual Saf Health Care,* 2008;**17**:216–223.

56. Von Laue NC, Schwappach DLB, Koeck CM. The epidemiology of medical errors: A review of the literature. *Wien Klin Wochenschr,* 2003;**115**(10):318–325.

57. Hickner J, Graham DG, Elder NC, Brandt E, Emsermann CB, Dovey S, et al. Testing process errors and their harms and consequences reported from family medicine practices: a study of the American Academy of Family Physicians National Research Network. *Qual Saf Health Care,* 2008;**17**:194–200.

58. Shojania KG. The frustrating case of incident-reporting systems. *Qual Saf Health Care,* 2008;**17**(6):400–402.

59. Hoffmann B, Beyer M, Rohe J, Genischen J, Gerlach FM. "Every error counts:" a web-based incident reporting and learning system for general practice. *Qual Saf Health Care,* 2008;**17**:307–312.

60. Dean B, Schachter M, Vincent C, Barber N. Prescribing errors in hospital inpatients: their incidence and clinical significance. *Qual Saf Health Care,* 2002;**11**:340–344.

61. Brennan TA, Leape LL, Laird NM, Hebert L, Localio AR, Lawthers AG et al. Incidence of adverse events and negligence in hospitalized patients: results of the Harvard Medical Practice Study I. *N Engl J Med,* 1991;**324**(6):370–376.

62. Leape LL, Brennan TA, Laird N, Lawthers AG, Localio AR, Barnes BA, et al. The nature of adverse events in hospitalized patients: results of the Harvard Medical Practice Study II. *N Engl J Med,* 1991;**324**(6):377–384.

63. Thomas EJ, Studdert DM, Burstin HR, Orav EJ, Zeena T, Williams EJ, et al. Incidence and Types of Adverse Events and Negligent Care in Utah and Colorado. *Med Care,*1999;**38**(3):261–271.

64. Studdert DM, Brennan TA, Thomas EJ. What have we learned since the Harvard Medical Practice Study? Chapter 1 (pp. 3–33) in: *Medical Error: What Do We Know? What Do We Do?* Edited by MM Rosenthal, KM Sutcliffe. San Francisco, CA: Jossey-Bass, a Wiley Company, 2002.

65. Baker GR, Norton PG, Flintoft V, Blais R, Brown A, Cox J, et al. The Canadian Adverse Event Study: the incidence of adverse events among hospital patients in Canada. *CMAJ,* 2004;**170**(11):1678–1686.

66. Gagnon L. Medical error affects nearly 25% of Canadians. *CMAJ,* 2004;**171**(2):123.

67. Otero P, Leyton A, Mariani G, Ceriani Cernadas JM, and the Patient Safety Committee. Medication errors in pediatric patients: Prevalence and results of a prevention program. *Pediatrics,* 2008;**132**(3):e737–e743.

68. Sinclair A. On the net. Medical error and patient safety. *CMAJ,* 2001;**165**(8):1085.

69. Sandars J, Esmail A. The frequency and nature of medical error in primary care: understanding the diversity across studies. *Fam Pract,* 2003; **20**:231–236.

70. Rosser W, Dovey S, Bordman R, White D, Crighton E, Drummond N. Medical errors in primary care. Results of an international study of family practice. *Can Fam Physician,* 2005;**51**:387–392.

71. Rasmussen J. Human errors: A taxonomy for describing human malfunction in industrial installations. *J Occup Accidents,* 1982;**4**:311–333.

72. Facione PA. Training the discovery of evidence of critical thinking. Pp. 279–285 (The Holistic Critical Thinking Scoring Rubric, p. 281) in: Facione NC, Facione PA. *Critical Thinking and Clinical Reasoning in the Health Sciences. An International Multidisciplinary Teaching Anthology.* Milbrae: California Academic Press, LLC, 2008.
73. Karlsen KA, Hendrix TJ, O'Malley M. Medical error reporting in America: A changing landscape. *Q Manage Health Care*, 2009;**18**(1):59–70.
74. Murphy JFA. Root cause analysis of medical error. *Irish Med J (IMJ)*, 2009;**102**(1):36.

Chapter 5

Analyzing Medical Error and Harm: Searching for Their Causes and Consequences

Executive Summary

Beyond descriptions of medical error events, identifying their causes or consequences requires two complementary strategies. The first is the a priori causal attribution model for detecting known or "obvious" causes like failures of tools or technologies. The other is the epidemiological model of causal proof requiring some kind of analytical observational or experimental methodology based on fundamental, field, and clinical epidemiology experience and expertise. Why is this the case? Medical error can be viewed as a "disease" with its own origins, causes, spread, and means to control or prevent it.

Explaining causes of a single medical error event or of very few events cannot be done by observational analytical methods as known to epidemiology. Causal attributions from single or few causes in exceptional situations can be studied like causes in clinical pharmacology (adverse effects) or in some proposals made for homeopathic medicine. For known causes, root analysis, for example, may be attempted although its results are not the proof of causes themselves. System analysis works in a similar direction. They are all a meticulous account of events and a generator of causal hypotheses yet to be demonstrated in most cases rather than a discovery of "new," still unknown causes.

There are two main ways to study causal relationships in the medical error and harm domain: (1) the epidemiological approach to discovering new and unknown causes; and (2) the approach involving "examining more or less known or 'obvious' causes" to try to discover them again as already known phenomena by detailing complex processes in time and space (webs) in complex systems of medical care and supporting organizations.

Cause–effect relationships in medical lathology may focus on single causes of medical error or causal relationships within the webs of causes or webs of consequences in a unifactorial or mutifactorial design. In any case, we want to demonstrate beyond randomness the existence of a cause–effect link between medical error and its causes or consequences in a proper cause–effect sequence that is sound, specific, consistent, plausible, and coherent with prevalent knowledge, confirmed by experimentation (if possible) and that satisfies the other casual assumptions, prerequisites, and proper criteria as still broadly discussed and expanded across current and historical epidemiological experience.

In research unsupported by causality, causal links are derived from up-to-date knowledge, analogies, various plausibilities (e.g., biological, technological, social, system- or actor-related, cultural, physical, chemical), consistency of findings, compatibility with prevalent standards of all kinds, and extrapolations from other domains of error, the rest being simply intuitive. The declaration of the nature of cause–effect relationships with all the strengths and weaknesses of the methods used in lathology then becomes mandatory.

Experimental methods of demonstration of cause–effect links in lathology are limited and most often impossible due to ethical reasons and a fortunate rarity of error events. As alternatives, epidemiological modeling or "thought experiments" (e.g., *what if* forms of reasoning) may be considered in the future.

In Chapter 6, we will look beyond the system paradigm of error to the human factor and, specifically, to the operating individual's reasoning and decision making that may be responsible for medical error events.

Thoughts to Think About

> The things related to surgery are the patient; the operator; the assistants; the instruments; the light; where and how; how many things, and how; the body, and the instruments; the time; the manner; the place.

—Hippocrates, 460–377 BC

We are too much accustomed to attribute to a single cause that which is the product of several, and the majority of our controversies come from that.

—Justus von Liebig, 1803–1873

When men are unable to form an idea of unknown and distant things, they judge them by what is familiar at hand. This axiom explains the inexhaustible source of all errors about the principles of human nature. These errors are embraced by entire nations and by scholars.

—Giovanni Battista Vico (Axiom 2, Principles of New Science Concerning the Common Nature of Nations, 1720–1725)

It takes less time to do a thing right than it does to explain why you did it wrong.

—Henry Wadsworth Longfellow, 1807–1882

The invalid assumption that correlation implies cause is probably among the two or three most serious and common errors in human reasoning.

—Stephen Jay Gould, 1941–2002

When everyone is wrong, everyone is right.

—Pierre-Claude Nivelle de la Chaussée, 1744

Irrationally held truths may be more harmful than reasoned errors.

—Thomas Huxley, 1825–1895

Man is made for science; he reasons from effects to causes and from causes to effects; but he does not always reason without error. In reasoning, therefore, from appearances which are particular, care must be taken in how we generalize; we should be cautious not to attribute to nature, laws which may perhaps be only of our own invention.

—James Hutton, 1788

Without knowing demonstrated causes of medical error and its consequences, its prevention and control are only an act of reasoned faith, however close to reality it might be. Either, we accept the causes of error as obvious and granted given our past experience, or we must identify causes of error from scratch as we have done with smoking and health or radiation and cancer. The error and harm cause–effect relationship is more challenging than disease causes. What would YOU do?

Introductory Comments

Our best knowledge of causes of medical error and medical error as a cause of what follows it (error outcomes) is a cornerstone of our likelihood of preventing error and avoiding disease outcomes that may be attributed to medical error. Hence, five views of error analysis and its impact are worthy of attention.

First, two cause–effect relationships are of interest:

- In linking medical error to its causes, medical error is a *consequence* of some occurrence.
- In linking medical error as a cause of some undesirable occurrence in a patient's health and disease course, medical error is a *cause* of some occurrence.

Second, any cause in lathology across the whole human experience falls under one of two methods of error causes and consequences recognition:

- We designate a priori and more or less arbitrarily, based on our expertise, events lived and other past experiences that this factor is a "cause" of medical error. Current experience with error at large operates this way. Are all these events, causes, and effects always as "obvious" as they often appear to us? We do not think so.
- We must accept that we need to demonstrate a "new," still poorly known or unknown cause–error association "from scratch" to build a more formal and stronger causal proof in waiting. That's what we prefer in epidemiology when faced with the necessity of explaining new associations.

Third, our choices of approaches to error etiology will largely be defined by two types of events:

- Errors are rather unique events. Their analysis is based on an a priori causal attribution. Various methods of single-case description and analysis apply; epidemiological methods do not. Life-threatening events and their potential causes, their interpretation, and attempts to control them cannot wait. For example, some toxic reactions in clinical pharmacology fall into this category.

- Errors occur in series that call for a formal epidemiological description and analysis, given their sheer numbers. Needle-prick injuries in hospital-based health professionals are good candidates for this kind of hospital epidemiology.

Fourth, there are two possible sources of confusion when discussing "error:"

- We consider error as a faulty way of (or absence of) reasoning and decision making. Under the constraints of time and need for action, not thinking about consequences, we do not hesitate to run down the staircase to a lower hospital floor and sprain an ankle.
- Error is still often confused across the literature and current experience with its consequences, outcomes, and the degree of their severity. In this case, such happenings are considered unavoidable consequences of error: an ankle sprain is an "error" caused by choosing to run down a slippery staircase.

Finally, we can consider error and patient safety as follows:

- If we accept that patient safety means freedom of error, solving the error problem will warrant patient safety.
- If we accept that patient safety depends not only on the freedom from error in healthcare but also on some additional important factors, medical error problem solving must not be entirely confused with patient safety itself.

Our thinking and decision making about errors in latholology will depend on and will be defined by the five aforementioned views and considerations: (1) cause or consequence; (2) error recognition; (3) frequency and urgency of error situations; (4) possible confusion between error and its outcomes; and (5) interchangeability of absence of error and patient safety. Our choices must always be specified and explicit in these contexts.

Searching for "New" (Not Yet Known) Causes and Consequences of Medical Error and Harm: Etiological Research, Analytical Observational Epidemiology

Causes in the domain of error represent a much wider set of entities than in other domains of human pathology like smoking and health or antibiotic or antiviral effectiveness in the treatment of infections. Webs of causes in latholology, if considered, must include a much wider array of potential or real causes not only from human biology but also from psychology, ergonomics, operational research, or any other field related to medical error. Figure 5.1 illustrates this case.[1]

Analytical epidemiology, which is the search for causes by observational methods like case control or cohort studies, remains equally challenging for the same previously mentioned reasons: definitions and taxonomy. Finding appropriate control groups, recording linkage, and gathering information from multiple sources, sites of activity, and other environments can sometimes be difficult, if not impossible. Consequently, presumed causes are most often declared authoritatively on the basis of past experience, research, and relational plausibility before any analysis is performed. Such a "causal attribution model"[2] currently prevails in medical error research and management.

Another gap to fill in current research on causes in the medical error domain is the fact that it still does not make always enough of a distinction between cause–effect relationships in which error is considered an effect or consequence of the actions of some causal factors producing error. In other words, error in a particular case is a dependent variable; its presumed causes are independent variables in the model under study. Other causal research focuses on the consequences of the medical error. In this case, error is an independent variable as a potential cause of some undesirable outcome in medical care. Outcomes (dependent variables) are consequences of the error that led to them. Moreover, taxonomies of error and variables under study are not always clear about their role in different models of analysis of different causal relationships.

Several types of cause–effect relationship in lathology may be of interest:

- Presumed factors, events, or other ("primary") errors leading to the error of interest in a unifactorial (single-cause) or multifactorial (more-than-one-cause) design.
- Possible causes of primary errors leading to the error of interest.
- Error as a cause of its one or more presumed consequences (accidents, outcomes, harm) in a unifactorial or multifactorial design.
- Causal relationships within webs of causes.
- Causal relationships within webs of consequences.

Whether some cause–effect relationship is studied by observation (in more or less controlled conditions) or by experimentation, causal analysis relies on comparisons between two or more series of observations. Figure 5.2 is an illustration of an epidemiological or structural path leading to the demonstration of a causal factor.

Similar attention like in this type of prospective or cohort approach applies to case-control studies based on comparisons of exposure to some presumed causal factor in the group of cases compared with the group of noncases.

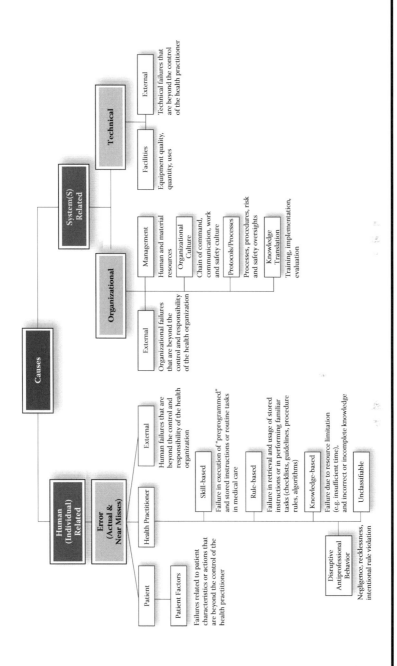

Figure 5.1 Classification of causes. (Modified from Chang, A. et al., *Int. J. Qual. Health Care,* **17, 2, 2005.[1])**

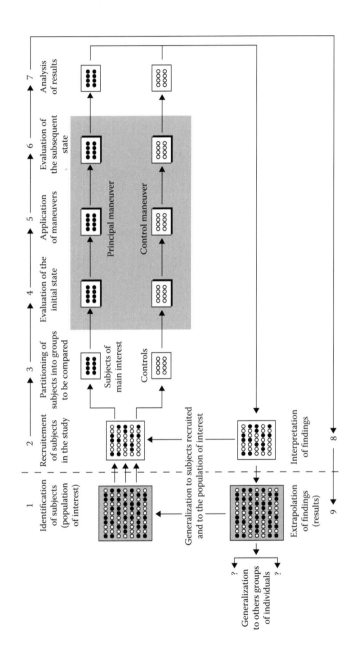

Figure 5.2 Epidemiological or structural path leading to the demonstration of a causal factor. (Modified and redrawn from Jenicek, M., *Foundations of Evidence-Based Medicine*, Parthenon Publishing Group Inc. (CRC Press), Boca Raton, FL, 2003.[3])

The quality of such demonstrations of causality does not depend only on how many causal criteria might be confirmed by the study, such as producing measures of strength of association (odds ratios, relative risks) or its specificity (e.g., etiological fractions, attributable risk percents). The quality of gathered data and comparisons of groups, their exposure, and rates of events at any step of the analytical process, for example, also define the meaning of the ensuing conclusions about causality. Clinical and fundamental epidemiology and biostatistics and evidence-based methodology in health sciences address in more detail these challenges,[3] as they apply equally to the medical error problem.

This situation of various frequencies of errors, different taxonomies of dependent and independent variables under study, and the quality and completeness of information about error happenings means that alternative methods of study to the epidemiological, mostly quantitative approach must be considered, including case studies in qualitative research, or mixed-method research methodology. Given the nature of the error and its frequent uniqueness, gravity, and ensuing urgency for solutions, such alternatives to classical analytical (etiological) research may often be the only feasible choices.

Moreover, gold standard cause–effect criteria of causation expanded from the philosophies of John Stuart Mill, Austin Bradford Hill, Brian McMahon, and Thomas Pugh and the U.S. Surgeon General as recently reviewed again[3] cannot always be applied given the qualitative and quantitative nature of error data and incidents (presumed causes) possibly related to them. For example, some more frequent events—consequences of error like needle-prick injuries in health personnel or falls of elderly patients in institutional care—may be an easier domain for etiological research than some more exceptional events like the removal of incorrect body organs or extremities.

In "classical" observational research of an epidemiological nature, we perform case-control or cohort studies and some of their hybrid designs to see if their results confirm or reject cause–effect criteria as summarized in Table 5.1.

If basic assumptions or prerequisites for causal consideration are fulfilled, we need to see if all proper causal criteria are confirmed or rejected one by one by the available research findings. Some criteria may be quantified as is the case of the strength of an association (by relative risks or odds ratios) or its specificity (exclusivity to a particular cause within a web of causes) in terms of etiological or preventable quantifications of one presumed cause compared with others (risk difference, attributable risk, attributable risk percent). Other criteria rely on judgment or on already acquired experience like biological plausibility. Our conclusions from observational research are confirmed, strengthened, or rejected by further experimental research and a systematic review of findings and evidence from the overall and best array of original

Table 5.1 Fundamental Prerequisites and Assessment Criteria of the Cause–Effect Relationship with Adaptations for the Error and Harm Problem in Medicine

Assumptions or Prerequisites (What Should Be Confirmed before Any Causal Criteria Apply)
• Exclusion of **randomness**. • Consistency of results with **prediction**. • Even observational studies respect as much as possible the same logic and similar precautions as used in **experimental research** • Studies are based on **clinimetrically valid data**. • Data are subject to **unbiased observations, comparisons, and analysis**. • **Uncontrollable and uninterpretable factors** are ideally absent from the study.
Criteria of Causation

Major (to be evaluated one by one)
• **Temporality** ("cart behind the horse")
• **Strength** (relative risk, odds ratio, hazard ratio)
• **Specificity** (exclusivity or predominance of an observation)
  – **Manifestational** ("unique" pattern of clinical spectrum and gradient as a presumed consequence of exposure)
  – **Causal** (attributable risk, etiological fraction, attributable risk percent, attributable hazard, proportional hazard)
• **Biological gradient** (more exposure = stronger association)
• **Consistency** (assessment of homogeneity of findings across studies, settings, time, place, and people)
• **Biological plausibility** (explanation of underlying mechanisms or nature of association)
Conditional (not always necessary for new discoveries)
• **Coherence with prevalent knowledge**
• **Analogy**

Reference (within the framework of the scientific method) • **Experimental proof** (preventability, curability): Clinical trials, other kinds of controlled experiments, or "cessation studies"

Table 5.1 Fundamental Prerequisites and Assessment Criteria of the Cause–Effect Relationship with Adaptations for the Error and Harm Problem in Medicine (*Continued*)

Confirmation (providing a clear research question, variables, and target population)
• **Systematic review** and **meta-analysis** of evidence

Source: Adapted with modifications from Jenicek, M., *Foundations of Evidence-Based Medicine,* Parthenon Publishing Group Inc. (CRC Press), Boca Raton, FL, 2003.[3]

studies, trials, and fundamental research. Distinctions between causality in the field of risk and prognosis are made; causes may be different. More information about causality in health sciences can be found in summarized and reviewed epidemiological literature.[2]

Originally, in the domain of infectious diseases, Evans-Koch postulates of causality sided with other causal criteria of the day. Can we also try to transpose Evans-Koch postulates in their current version[4] and apply them to the error problem?

1. Prevalence of disease (occurrence of error) should be significantly higher in subjects exposed to the presumed causal factor (individual or "direct" or system factors) than in unexposed controls.
2. Exposure should be more frequent among those with the disease (cases or subjects of error) than in nondisease (nonerror) cases.
3. Incidence of error is higher in those exposed to the presumed cause of error compared with unexposed individuals.
4. The log-normal distribution of incubation period criterion does not apply here.
5. The biological gradient of severity (frequency and seriousness of error and its consequences) is rather exceptional knowledge in the error domain.
6. A measurable host response (error events) following exposure (to the presumed cause or causes of error) should have a high probability of appearance in those lacking this prior exposure.
7. Experimental reproduction criteria and consideration do not apply here for ethical reasons.
8. Modification of the cause should be followed by modification of the incidence of error.
9. Prevention or modification of the host's response (responses of individuals or systems to error by, e.g., training, education) on exposure to its

presumed causes should decrease or eliminate the disease (error and its consequences).

10. All of the above should make sense in view of the epidemiological, social, ergonomical, technological, environmental, or operational characteristics of a specific error problem.

Should we also look at causes of error in light of such causal criteria? We should always try.

Currently, the National Center for Patient Safety of the U.S. Department of Veteran Affairs, proposes five rules of causation in the patient safety domain:[5]

1. Causal statements must clearly show the "cause and effect" relationship.
2. Negative descriptors (e.g., poorly, inadequate) are not used in causal statements.
3. Each human error must have a preceding cause. (This criterion overlaps the first.)
4. Each procedural deviation must have a preceding cause.
5. Failure to act is causal only when there was a preexisting duty to act.

Those rules should be considered in addition to already customary causal criteria in epidemiology.

Error in medicine is another health phenomenon that is, in most cases, subject to similar considerations as any other aspect of health, disease, and their management. There is no way other than searching for causes by other methods, however less satisfactory in the error domain they might be.

Ideally, in the spirit of Table 5.1, we want to demonstrate beyond randomness the existence of a cause–effect link between medical error and its causes in a proper cause–effect sequence that is sound, specific, consistent, plausible, and coherent with prevalent knowledge, that is confirmed by experimentation (if possible), and that satisfies the other causal assumptions, prerequisites, and proper criteria as still broadly discussed and expanded across the epidemiological literature.

Is there any other alternative to classical cause–effect research in the medical error domain? Yes, but no matter how different it is from the golden standard, we should not feel guilty about concessions. The nature of the error problem in medicine requires some flexibility, adaptations, and refinements.

Epidemiological analysis of cause–effect relationships in lathology is possible and necessary if sufficient numbers of cases of error and harm are warranted and available. An epidemiologist may say, less seriously, "Give me a hundred or hundreds of cases of error and harm, and I will find them a good control group and produce a nice case-control study of some cause–effect relationship in this case." Or, he or she may also suggest, "Do you have thousands or hundreds of thousands of individuals at risk of being exposed to the risk of medical error and harm during a specified, long enough period of time? If so, I will find you a suitable control group of individuals unexposed to such risk and produce for you a nice cohort study demonstrating a cause–effect relationship between error and harm as a cause or consequence of something."

Fortunately for patients and to the despair of epidemiologists, numbers of cases of medical error and harm do not always warrant enough a valid epidemiological study. In these situations, other ways to study cause–effect are the only alternatives. The next sections of this chapter summarize current experience in this "epidemiologically alternative" domain.

Challenge of Deriving Cause–Effect Relationships from One or Very Few Observations: An A Priori Causal Attribution

This situation puzzles many researchers in the health domain. Classical criteria of causality do not seem to apply or cannot be confirmed from the available number of cases and related information. What are the alternatives?

Challenges of Limited Causal Proof or Causes Yet to Be Established

What can we do when data and information about error are limited, impossible to obtain, not yet available, or simply inaccessible for ethical and other reasons?

In the case of research-unsupported causality, cause–effect links are derived from the following:

- Associations based on past and present *knowledge* (e.g., books, articles).
- Associations based on *analogy,* which is based itself on recollection and comparisons with past or present experience.
- *Biological, social, technological, system-functional, interactional, cultural, physical, chemical,* or other *plausibility* as seen through experience in relation to the subject of the cause–effect study. (Biological plausibility is probably not enough in the error domain.)
- *Coherence with prevalent knowledge.*
- *Consistency of findings* across published and otherwise reported and recorded past and present experience.

- *Compatibility* of our views *with prevalent standards* (norms, guidelines, position statements, management orders, however questionable they might be).
- Various *extrapolations* even if based on some kind of systematic review of past cases and experience.
- Some other cause–effect relationships are purely *intuitive*, waiting for hypotheses generation and formal proof.

Each of these alternatives to the more complete and formal etiological research is worth their worth, but very often this is the only information available before better proofs become available. We have no other choice than to deal with them, to judge them by their own and right value (with all inherent limitations), and to make our conclusions as sound as the available information requires. It is not shameful to report critically causal considerations that are still incomplete or imperfect provided that such limitations are critically overviewed in the context of the author's error research and that his or her conclusions are not exaggerated.

To make any of our causal statements more explicit in the medical error and harm domain, it should be mandatory to state the nature of the "cause" or "causal factor:" how and why was it recognized as a "cause?" Was it based on intuition, experience, analogy, extrapolation, a formal causal proof stemming from current causal criteria, or was something else involved?

Some of the alternatives and challenges to causal proof in the error domain certainly deserve consideration.

Is It Possible to Estimate and Analyze Probabilities of Rare Events?

Let us suppose that we want to compare medication errors and their consequences, which are even rarer than nosocomial infections. They occur in two hospital environments where medical, nursing, and pharmacy teams communicate, interact, and proceed differently in treatment decisions and actions.

For example, estimation of probabilities of rare events, proposed by Alemi[6,7] in the framework of probabilistic risk analysis, might prove of particular interest in the medical error and harm problem domain. In some situations, epidemiologists cannot use, whatever the reason may be, conventional epidemiological measures of risk such as incidence rates or other expressions of probability that an event will occur within a certain, often shorter period of time. From the rare events methods available and experience with probabilistic risk analysis (discussed later in this chapter) acquired in the space industry (shuttle flights), nuclear industry (reactor safety), terrorist attacks analyses, natural disasters (floods, coastal events, earthquake predictions), environmental pollution and environmental health, or waste disposal[6,7] originate rare events probability assessment methods: fault trees, similarity judgments, importance sampling, and time to the event. The latter is one of several approaches added to an already well-established time-to-event study methodology developed by biostatisticians in the domain of clinical trials, prognosis, and survival analyses.[8]

In the case that the event either happens or does not, has a constant probability of occurrence, and the probability of the event does not depend on prior occurrences of the event, assuming that the sentinel event has a geometric distribution, the probability of a rare event, p, can be estimated from the average time to the event by using the formula

$$p = \frac{1}{1+t}$$

where p is the daily or per visit (or another event of care in health sciences) probability of sentinel event, and t is the number of days to or the number of visits (or other acts in clinical and other care).

In Alemi's example of calculating the daily probability of medication errors, the last 10 incidences (sentinel events) are identified and the days between two consecutive incidences (sentinel events) are calculated and averaged yielding the average time to reoccurrence to estimate the daily probability of such medication error.

In the case of an event occurring:

- Once per year or less (deemed low occurrence), a calculation of 1/(364 + 1) yields the rare event probability of 0.0027.
- Once per six months or less (deemed medium occurrence), a calculation of 1/(6 × 30 + 1) yields the rare probability of 0.0056.
- Once per week or more (deemed very high occurrence), a calculation of 1/(6 + 1) yields the rare event probability of 0.1429.
- An occurrence of one per day or more (deemed extreme) yields a 1/1 calculation of a rare probability of 1.

Whether such estimations and rare probabilities can be used instead of conventional epidemiological estimations of risk (derived from observations of more frequent events based on incidence rates) to evaluate the strength or specificity of causal associations (relative risks, etiological fractions, attributable risks, attributable risk percents) remains to be seen. The current literature is not rich in these matters. Promising or not, so far do we have anything better in the case of studies of occurrence and causes of rare events?

Table 5.2 States Indicative of Causality that May Apply (but Not Always) to Unique, Single, or Infrequent Events in the Domains of Medical Error and Harm

Adverse Effects in Clinical Pharmacology[9]	*Single-Case Causality Assessment and Criteria in Homeopathy and Alternative Medicines*[10,11]
• Extracellular or intracellular tissue deposition of the drug or metabolite • Specific anatomical location or pattern of injury • Physiological dysfunction or direct tissue damage that can be proved by physiochemical testing • Infection as a result of the administration of a potentially infective agent or because of demonstrable contamination	• Time-figure correspondence • Space-figure correspondence • Morphologic pictorial correspondence • Intensity-figure correspondence • Drug-picture correspondence • Therapeutic idea correspondence • Ping-pong correspondence

Single-Error Event or Few Error Events Reporting

Can single-case reports or case series be a proof of causality? In some cases, perhaps.

Table 5.2 presents a list of considerations of causality beyond customary criteria in epidemiology as proposed for adverse effects studies in clinical pharmacology[9] and effectiveness of homeopathic medicine requirements[10] and reviewed also elsewhere.[11] In clinical pharmacology, criteria rely considerably on localization of drugs, metabolites, and plausible anatomical and physiological responses. In homeopathy,[10] the proposed criteria focus on the rapid, if not immediate, biological response to stimuli and the similarity (or explainability based on some past experience) of site, time, space, pathological physiological function, or dose-effect associations ("shapes," "correspondences," "fits" to their authors).

Offbeat Searches for Causes: Siding with Mainstream Epidemiological Experience

Struggling with the frequency of error, the vast set of domains from which its causes originate and the extended web of error causes itself led lathologists to consider approaches other than formal analytical observational or experimental methodologies.

Root Cause Analysis in the Health Domain

An example of an error specific approach may be the root cause concept and analysis (RCA). It may, however, be driven to the identification of a single cause within a web or linear sequence of causes. The introduction of the RCA principles in health sciences is attributed to Bagian et al.[12] in 1992, updated and with related items in the *VHA National Patient Safety Improvement Handbook*.[13]

Root cause analysis as also briefly mentioned in Chapter 2 in the context of general error was subsequently used in the health domain. Its origins are in psychology where it was developed to identify the basic and causal factors that underlie variation in performance. It is not a single and sharply defined methodology, being possibly safety based, production based, process based, failure based, or system based.[14] In the medical error context, it aims essentially for a better understanding at the system level. Elsewhere, it was adopted by the manufacturing industry and adapted to mitigate production errors and their consequences. It is an event analysis tool applied retrospectively to identify, evaluate, and understand adverse, latent, sentinel, and operator factors in the development and happening of error. Root cause analysis aims to answer at least in part three important questions:

1. What happened?
2. Why did it happen?
3. What might and should be done about it?

Hence, its ambitious objective is the description, analysis of causes, and identification of preventive and control measures, all of which are usually treated separately in epidemiological research.

Root cause analysis in our domain is defined as "a structured process for identifying the causal or contributing factors underlying adverse events or other critical incidents.[15]" The key advantage of RCA over traditional clinical case reviews is that it follows a pre-defined protocol for identifying specific contributing factors in various causal categories (e.g. personnel, training, equipment, protocols, scheduling) rather than attributing the incident to the first error one finds or to preconceived notions investigators might have about the case.[15]

It is also characterized and defined by Runciman et al.[16] as "a systematic iterative process whereby the factors that contribute to an incident are identified by reconstructing the sequence of events and repeatedly asking why until the underlying root causes have been elucidated."

Croskerry et al.[17] define the RCA as "an analytic tool that can be used to perform a comprehensive, system-based review of critical incidents. It includes the identification of the root and contributory factors, identification of risk reduction strategies, and development of action plans along with measurement strategies, to evaluate the effectiveness of the plans."

Series of "why" questions and answers to them move from one potential causal factor to another before the incident aims at the establishment of a mostly linear model and sequence of events that precede the incident, the root cause representing the event that started it all. Conceptually, root causes fall into the category of necessary causes (conditions) rather than sufficient causes.[4,18]

A root cause itself is, let us remember, an initiating cause of a causal chain that leads to the outcome of effect of interest, an element where an intervention could reasonably be implemented to change performance or prevent an undesirable outcome.[19] In simplest terms, we may see it as some kind of "manifestation of error."

Proponents of root cause analysis are well aware of its role in the causal research of error: though the definition of RCA emphasizes analysis, the single most important product of RCA is descriptive—a detailed account of events that led up to the incident.[15]

RCA has for its declared purpose:[20]

■ To determine human and other factors involved in critical incidents.
■ To determine related processes and systems.
■ To analyze underlying causes and effect systems through a series of "why" questions.
■ To identify possible risks and their potential contributions.
■ To determine a potential improvement in processes and systems.

The root cause analysis and control process consists of several steps:[21]

1. Acquisition of an initial understanding of the problem based on facts known in time.
2. Gathering of additional information relevant to the critical incident.
3. Literature review (review of past experience).
4. Establishment of a detailed picture of a sequence and interaction of human and system factors leading to the incident.
5. Determination of contributing factors and root causes.
6. Identification of additional (incidental) findings, not necessarily related to the incident but relevant to patient safety or to patient care in general.
7. Formulation of causal statements (based on the previous steps).
8. Development of actions to correct the problem and propose ways to prevent its occurrence in the future.
9. Implementation of the action plan.
10. Measurement and evaluation of the effectiveness of actions.
11. Sharing results of improvement.

An example of a detailed root cause analysis describes and analyzes a fluo-rouracil incident involving the fatal administration of fluorouracil to a cancer patient.[22] Instead of being administered over four days, it was administered over four hours. This example shows both major contributions of RCA to the under-standing and management of the error event and its major weakness from an epidemiological point of view.

A 43-year-old woman patient received, as part of her radio- and chemo-therapy for an advanced nasopharyngeal carcinoma, fluorouracil, and cisplatin in an ambulatory setting. It was found one hour after the drug administra-tion that the infusion of fluoprouracil, which should have been done over four days, was administered over four hours. Four days later, the patient was admitted for mucositis and pancytopenia, experienced hemodynamic collapse and multiorgan failure before dying due to cumulated fluorouracil and cispla-tin toxicity.

As an initial impression and understanding before root analysis, an error due to a discrepancy between the treatment order and administration instructions was attributed to a nurse in charge of programming an administration pump according to the order and administration instructions, an action (incorrect rate of the chemotherapy pump) considered to have led to this critical incident.

The event was then subject to root analysis based on a detailed and rigorous chronological portrait of events, actions, and manifestations both in the health professionals' decisions and actions and patient responses between the onset of the current ambulatory care and the death of the patient. After the event analysis, illustrated here in Figure 5.3, several additional potential elements of system activities and functioning were proposed besides the fluorouracil over-dose as components of final understanding of this critical incident: design of the chemotherapy protocol, inability to mitigate harm from an unknown maxi-mum tolerated or lethal dose of this kind of chemotherapy, programming of and possible auto-correction of the pump device, labeling of information moving between the pharmacy and nurses, missing pump programming directions in medical records and orders, lack of communication and feedback, low knowl-edge of hazards, complex workload, and multitasking and other important asso-ciated findings.

Ensuing proposals for future corrective measures includes both the preven-tion of human (operator error, active error) error at the "sharp end" of the event like executing orders in pump programming and drug administration and pos-sible latent factors corrections like chemotherapy administration pump design to prevent errors in orders or organization of overall hospital care practices by nurses, pharmacists, and other health professionals.

As in any other method and technique, root analysis shows both its strong and weak sides. On the strong side:

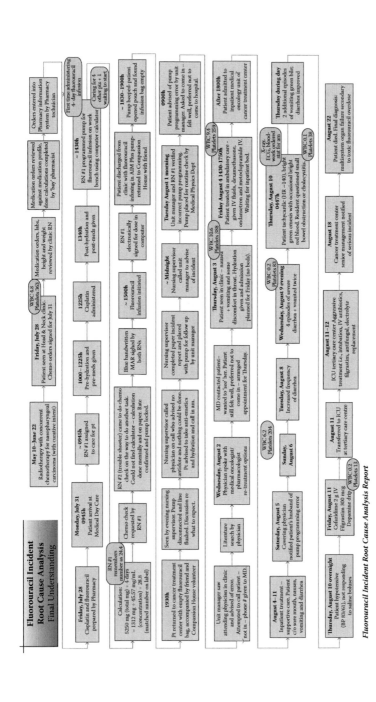

Figure 5.3 Chronology of events in the development of a fluorouracil treatment critical incident as a basis for root analysis and corrective measures. (From Institute for Safe Medication Practices Canada, analyzing team, *Fluorouracil Incident Root Cause Analysis,* formatted for Web posting May 22, 2007. With permission.[22])

- It establishes a detailed chronological picture of events, actions, and individuals involved in the event.
- The problem is placed within the context of past experience.
- May work better and show more useful in the analysis of unique cases.[23]

On the weak side:

- It establishes root causes only on the basis of past (literature and other) experience relying on the five RCA-style root causal criteria that are not causal proofs by themselves.
- RCA relies heavily and primarily on the analysis of a linear sequence of events. Even tree analysis (ramification of events in time)[24] does not replace a better view of the web of causes; interaction of possible causal factors is omitted.
- RCA does not always balance human, system, and environment (technology) factors.
- A retrospective hindsight analysis does not necessarily fit what might be seen as a reality and basis for future.[25]
- A single root cause is rather exceptional and may be misleading since most failures result from interactions between latent failures and direct action failures.[26]
- It may identify "necessary but not always sufficient" causes of the event.[27]
- It may be and is often a subject of post hoc ergo propter hoc fallacy.
- RCA is a loaded process, subject to hindsight bias and assigning the relevance to events according to already available knowledge.[28]
- RCA-related actions and outcomes are influenced by RCA inquirers' knowledge, interpersonal relationships, hierarchical tensions, biased perspectives, and preexisting agendas.[28]
- Analyses may be ended as soon as they lead to the most convenient root cause rather than real cause.[28]
- Clear actions resulting from RCA do not necessarily follow.
- RCA ensuing changes in process and outcomes of care may potentially create unsafe situations themselves.[28]
- Continuous changes to the process may drift activities away from their target.[28]
- Most studies do not measure resulting processes and outcomes.[28]
- In the health domain, RCA is difficult to organize and implement as a measure of sustainability and improvement.[29]
- The acquired RCA experience provides anecdotal evidence that RCA improves safety, rather than evidence of the effectiveness of implementation of its results, still to be improved.[29]

- Other methodological improvements must be considered like coupling RCA with human factors engineering[30] and structuring and standardizing the RCA process,[31,32] taking into account environmental, team, individual, patient, task, defenses, and other underlying factors.[33]
- True root cause (without which the incident would not occur or future incidents preventable by the elimination of the root cause) understanding and future management remain challenging for any epidemiologically minded person.
- A more detailed analysis of mistake mechanisms such as deficiency in logic, critical thinking, and decision making as the essence of error leading to critical incidents is more than limited in most cases.

Such strengths and weaknesses of RCA must be seen as strategies for handling the problem that is often rare and unsuitable due to this and other reasons for formal epidemiological cause–effect analysis. RCA may be the best way of causal analysis given the circumstances, but RCA conclusions about causes should not be emphasized beyond the scope of the RCA proof.

Other improvements should consist of better executed and better implemented RCA studies and an evaluation of their implementation and effect.[29] The essential requirements for a good root cause analysis have been worked up by the U.S. Department of Veteran Affairs National Center for Patient Safety.[34]

From an epidemiological standpoint, root cause analysis then is not a substitute for a formal cause–effect demonstration and proof! It is one of the bases on which we may better formulate our hypotheses about cause–effect relationships in the medical error domain. General rules of causation in epidemiology are summarized in the first section of this chapter, and some of them were applied to human error:[35]

- Causal statements must clearly show the "cause and effect" relationship.
- Negative descriptors (e.g., weak, inadequate) are not used in a causal statement.
- Each human error must have a preceding cause.
- Each procedural deviation must have a preceding cause.
- Failure to act is causal only when there was a preexisting duty to act.

Wald, Shojania, and others[36,37] conclude that root cause analysis as a research tool is limited by its retrospective and speculative nature and insufficient evidence about its effectiveness, efficacy, and efficiency (added by us) but that it may represent an important qualitative tool complementary to other techniques employed in error reduction and that it lends a formal structure to efforts to learn from past mistakes.

Are there any other approaches and methods to consider?

Other Approaches to Cause–Effect Studies in Lathology through Observational Methods

Recent initiatives in etiological research in the medical error and harm domain include methods and techniques that are extensions or complementary to root cause analyses or they stand by themselves as efforts to propose additional contributions to solving the cause–effect problem in this domain.

Causal Trees

In root cause analysis, as seen earlier in the chapter, a linear sequence of events may be built to trace flaws in the process of healthcare or practice.

Causal tree analysis is defined as "an investigation and analysis technique used to record and display, in a logical, tree-structured hierarchy, all the actions and conditions that were necessary and sufficient for a given consequence to have occurred."[37]

Causal trees represent an additional, graphic in substance technique of reconstructing a sequence of events both in a time and space relationship. The index harm (the main event) is placed on the top of the graphic representation from which a web of events, "down the tree in time" is built to its "lowest" flaw, a possible root cause of the main incident. Figure 5.4 illustrates in the form of a causal tree an incident analysis of a knife cut suffered by a laboratory aide.[38]

Weber[39] proposed that a quantitative causal tree analysis may prove useful in the detection of possible causal chains leading to harm. Based on his computer science experimentation, a controlled lesion method (CLM) should lead to the identification of causal tree structures and might prove necessary for a causal discovery system.

Elsewhere, causal trees were built to better understand possible causes of infection after surgery, giving the wrong medication to a patient, anesthesia error,[40] administering an outdated vaccine to a patient,[41] and analyzing incidence and causes of critical incidents in emergency departments.[42]

As with any other method of inquiry, causal factor tree analysis has its pros and cons. It provides a structure for the recording of evidence and display, but only for what is known. Wilson[43] proposes that, through application of logic checks, gaps in knowledge may be exposed. However, it cannot easily display time dependence, and complex sequences may be difficult to analyze. It shows where unknowns exist but provides no means of resolving them. Also, stopping points can be somewhat arbitrary.[43]

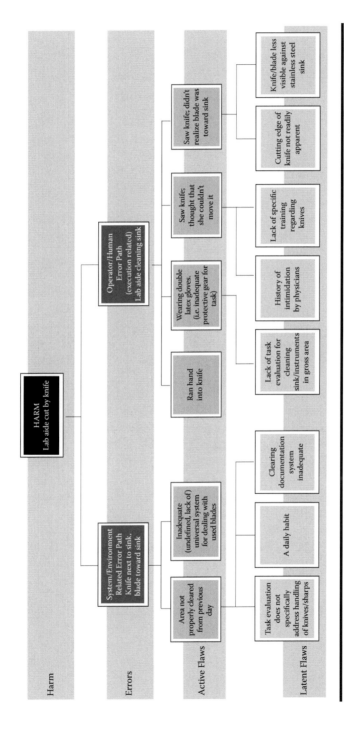

Figure 5.4 A causal tree example of a knife cut in a laboratory aide. (Redrawn with modifications from Williams, P. M., *Baylor Univ. Med. Center Proceedings*, 14, 2, 2001.)

Probabilistic Risk Analysis

In lathology, probabilistic risk and hazard modeling and analysis were also proposed,[6,7,44-48] originating in the aerospace industry and in the areas of nuclear safety, natural disasters, and terrorism.

As software-supported computational technique, *probabilistic risk analysis (PRA)*, represents a method of addressing the triple risk—what can go wrong, how likely is it, and what are the adverse consequences and how likely are they—as related to the performance of complex systems to understand likely outcomes, sensitivities, areas of importance, system interactions, and areas of uncertainty.[48] As additional methods become available, *human reliability analysis (HRA)* in the nonmedical domain deals with methods for modeling human failure, and *common-cause-failure analysis (CCF)* deals with methods for evaluating the effect of intersystem and intrasystem dependencies that tend to cause simultaneous failures and thus significant increases in overall risk.[49] PRA, as its name implies, introduces probabilities within the analysis of possible associations between adverse events and their potential causes.

PRA has its positive and negative sides:
On the positive side:

- It is based on some a priori knowledge of the realistic probability of occurrence of the adverse event.
- It is a consistent method of aggregating the risk of several events co-occurring.
- It can be used to combine the effect of a sequence of events in a consistent and logical fashion.

On the negative side:

- Rare probabilities cannot be estimated accurately.
- It may divert attention from the task of risk modification.
- Compared with other traditional risk analyses, focusing on a restricted list of risks, it may be faster, but it may not always provide a specific direction for change.
- If situations and conditions change, it is not relevant in predicting future events.[47]

PRA or analysis belongs to the field of *failure modes and effect analysis (FMEA)* as proposed also for uses in the health domain. In the aforementioned spirit, it consists of several steps:

1. Identifying the outcomes of interest.
2. Assembling fault trees.
3. Developing the model.
4. Adding probability estimates.
5. Improving the model.
6. Adding socio-technical components to the fault tree.
7. Analyzing and interpreting the model.

Marx and Slonim[44] conclude that PRA has advantages over FMA in that it considers multiple combinations of failures and allows identification of critical failure paths.

Applying probabilistic risk analysis to medical error, Hovor and O'Donnell[46] use available information (e.g., experts, literature, local experience) to predict probabilities of future potential sentinel events based on current knowledge of near misses of the different causes of medical error, such as transcription, medication, pharmacy dispensing, inventory and storage activities, and medical practices in the healthcare facility of interest in a Bayesian causal network model.

Alemi summarizes the criticism of his own PRA in three points:[47]

- To estimate rare probabilities cannot (or may not) be estimated accurately. "Time to an event" information uses should help.
- PRA is not practical and will divert attention and resources from the task of risk modification. Its further uses and evaluation should evaluate this point.
- History from which information is gathered and used in PRA is not relevant in predicting future events. In medicine, it might prove otherwise.

This type of analysis (PRA) is a prospective approach to the prediction of error and harm. As it may prove to be a tool to prevent undesirable events in healthcare, we will discuss it further in Chapter 6 in relation to prognosis and in Chapter 7 related to *healthcare failure mode and effect analysis (HFMEA)*.

Significant-Event Analysis

The field of general medical practice evaluation in the United Kingdom is among the first to introduce *significant-event analysis (SEA)* in the medical error and harm domain. Bowie et al.[50] define SEA (with our minor modification) as "a qualitative method of clinical audit based on a synthesis of traditional case review(s) and the research principles of the critical incident inquiry to perform a more in-depth, structured analysis of an event identified to be 'significant' by a healthcare team." It may be considered a "screening procedure" to identify cases for more complete analysis, prevention, and control. There are 27 SEA uses illustrating their point.

The Royal College of General Practitioners sees SEA as "a structured approach to reviewing events in any area of work that has occurred in general practice and family medicine: management of acute and chronic cases, prevention, organization and management, essentially in the form of an audit."[51]

It is essentially a qualitative inquiry into such events as prescribing errors, failures to act given some abnormal result, failure to refer, to deal with an emergency call, breaches of confidentiality or breakdowns in communication, death, myocardial infarction, and admissions.[52]

A variable number of steps is suggested across the available information:[6,51-54]

1. Choosing an event: What is a significant event? Why is it a significant event? What factors led to this event?
2. Collecting and collating the factual information.
3. Choosing a team to perform SEA: Who was, is, and will be involved in the happening, its current analysis, and its future management?
4. Preparing the team for a meeting and teamwork.
5. Running a structured analysis of the significant event at the meeting and analyzing and answering questions like, "What happened? Why did it happen? What was learned? What was changed?" In the spirit of this meeting and team experience, what actions need to be taken as a result of this event?
6. Monitoring progress of all actions that were agreed upon and implemented.
7. Seeking educational feedback and sharing the experience with other practitioners and other stakeholders in general and family practice and beyond (if appropriate). How was the event handled? How could it have been handled differently?
8. Proposing lessons that can be learned for future reference.

Other types of event analyses may follow.

Systems Analysis: Beyond Incident Reports and Root Cause Analysis

Seeing the weakness of the root cause analysis, including the oversimplification of error history by focusing on a single cause (or very few), the U.K. Clinical Safety Research Unit proposed a *system analysis* focusing on the broad system of events including a chain of contributing factors combined in the future with incident analysis in anticipation of future problems based on current experience.

Vincent et al.[55] under the auspices of the Clinical Risk Unit of the University College London and the Association of Litigation and Risk Management

(ALARM) suggest here that analyses of clinical incidents should focus on organizational factors rather than on individuals and use a formal protocol ensuring a systematic, comprehensive, and efficient investigation. Attention is paid to care management problems, clinical context and patient factors, and general and specific contributory factors.

The nine steps of their investigation process[53] are as follows:

1. Ascertainment of the incident.
2. Triggering the investigation procedure with proper notification.
3. Establishment of circumstances to decide which part of the process of care requires investigation.
4. Structured interview of the staff.
5. Repeating interviews if new management problems have emerged during interviews.
6. Collating interviews to perform a composite analysis under each care management problem identified and identifying related general and specific contributing factors. Examples of a care management problem include failures of observation and action, delays in diagnosis, incorrect risk assessment, faulty equipment, not carrying out preoperative checks, lack of supervision, wrong treatment, and wrong location surgeries.
7. Elaborating a report of events that includes recommendations or prevents recurrence.
8. Submitting the report to major stakeholders in the clinical incident domain under investigation.
9. Implementing actions arising from the report and monitoring the progress of the incident understanding, management, and control.[44,46]

Vincent et al.'s *incident analysis*,[55,56] like root cause analysis, is not devised to identify some new unknown causes in an epidemiological manner. "Causes" of incident are drawn from the system of care based on relevant experience already at hand and applied to the new situation (incident) under inquiry. Both root cause analysis and incident system analysis provide, however, a detailed picture, chronology, and sequence of events for further hypotheses generation and experience generated actions. They do not replace a more formal causal proof per se.

Neither of the aforementioned cause searches should exclude deficient omissions and uses of evidence of variable quality, completeness, and relevance in a fallacy-free reasoning and decision making made by the operator as a human factor at the end of the process and system leading to medical error and its consequences (harm). Chapter 7 offers some strategies, methods, and techniques to approach this challenge at the "sharp end" of error history.

Experimental Demonstration of Medical Error and Harm Causes and Its Compromises and Alternatives

As in any application of the scientific method to the evaluation of a cause–effect relationship beyond the observational domain, health inquirers themselves decide who will constitute the group of interest, who will be a control group, what kind of preventive or therapeutic interventions will be chosen and compared, how to make such analyses free of bias, and how realistic interpretations of findings should be.

Necessarily, the error phenomenon in medicine like any other mass health phenomenon such as disease and health itself should benefit from the epidemiological approach to its solution. Like many other entities, it varies in frequency, nature, and solvability from one activity, place, and human and environmental context in which various types of error occur to another.

Ethical considerations practically exclude the experimental method from the search for medical error causes.

No Experimentation or Observational Research Is Feasible? Thought Experiment ("What If" Reasoning) to the Rescue

Another alternative to classical causal research may be the thought experiment concept, a still untested conceptual territory in the error domain. A *thought experiment* (Ger. *Gedankenexperiment*, since 1820, or experiment conducted in the thoughts)[57] is a sort of *what if* reflection.[58] It is a proposal for an experiment (by way of an intellectual exercise) that would test a hypothesis or theory, but cannot actually be performed due to practical (e.g., means, ethics, feasibility) limitations. Instead, its purpose is to explore in a reflective way the potential consequences of the principle in question. In the error domain, experimental proofs are very limited given the most fundamental ethical considerations related to the error consequences manipulations. However, thought experiments may be attractive options in the future. Current experience with their methodology remains limited in the error domain for the moment.

Thought experiments are defined as devices of the imagination used to investigate the nature of things.[59] Their contribution to modern physics, quantum mechanics, relativity, or philosophy is much more spread than their role in health sciences.

In a thought experiment, data are formulated into a truth table where researchers ask themselves a question. If the answer is yes, another question is raised, and so on. Any "no" answer means going back to a preceding "yes" step and proceeding accordingly in a modified way.

Many philosophers prefer to consider thought experiments to be merely the use of a hypothetical scenario to help understand the way things actually are.[57] The thought experiment process is the process of employing imaginary situations to help understand the way things really are. It is a kind of trade-off between reality and its substitution in some situation where it is still feasible to evaluate given the circumstances. Thought experiments display a patterned way of thinking allowing us to explain, predict, or control events in a still more productive manner than by some of the haphazard alternative ways of handling the problem. In science, thought experiment is a "proxy" experiment.[57] However, wherever relevant, it must follow the same causal criteria as in any other observational analytical or experimental research. More about this can be found in the literature summarized elsewhere.[3,4]

Thought experiments, however, do not replace experimental proofs. Also, they are necessarily loaded with the knowledge and experience of the thought experimenter. Hindsight bias cannot be excluded. They may serve as a heuristic aid or provide support for a theory.[59] For Norton, as summarized by Brown,[59] any thought experiment may also be seen as a (possibly disguised) argument containing premises grounded in experience that lead either in an inductive or deductive way to a conclusion or claims.

Among many reasons for thought experiments,[57] let us quote extrapolations beyond boundaries of established facts, explaining past, predicting future, attributing causation, and ensuring future avoidance of past failures. We may expect them to provide us with some kind of a priori knowledge before formal (empirical science) experimentation is attempted (if feasible). But what else can we do if observational analytical research or experimentation is impossible whatever the limitations of such impossibilities might be? Medical error is one of these situations.

The thought experiment entry in *Wikipedia, the Free Encyclopedia*[57] also offers a solid selection of fundamental articles and monographs covering this topic that should not escape the reader's attention.

A Word about Modeling in Epidemiology and Lathology

Modeling or simulations in epidemiology may be considered a kind of thought experiment. *Models in epidemiology* are abstract representations of the relationship among logical, analytical, or empirical components of a system.[4] For example, epidemics may be seen as systems, that is, sets of interacting and interdependent entities forming an integrated whole and embodying a relationship that differentiates them from other sets. Epidemics and their components may be seen and studied as systems. Models may be used in simulations to approximate the functioning of the "real" model system in a kind of thought-, mathematical-, or computer-assisted experiment. Their uses in the medical error domain are so far more than limited. However, probabilistic risk modeling and analysis are also proposed.[6,7,44-48]

Simulations may prove useful well beyond military medicine[60] or clinical specialties like surgery or gynecology and obstetrics,[61] medical error and harm included.

Is the Mainstream Epidemiological Methodology of Causal Research Feasible in the Domain of Medical Error and Harm?

Not only is epidemiological research methodology in the causal domain feasible, but its uses also should be expanded for the benefit of our knowledge and control of the medical error domain. In addition, it is desirable and necessary if we want to go beyond the preconceived ideas of causes however reasonable, realistic, relevant, and otherwise attractive they may be at first sight.

Often, etiological research focuses on the harm caused by human or system error and not on error itself. This may be irrelevant in the case of sensory-motor errors like injuries caused by sharp objects in surgical specialties and general care. More discrepant findings may be expected in studies of medical errors produced by faulty reasoning and decision making associated or not to some questionable evidence and represented by the harm they produce under the label of, for example, outcomes, incidents, critical incidents, and near-miss studies. Researcher and reader beware!

Most available studies are based on some kind of *epidemiological surveillance* and reporting providing a generally better picture of the medical error problem in the sense of descriptive studies outlined in Chapter 4. Watterson's EPINet surveillance results[62] illustrate well this methodology and its results. Conclusions and recommendations stemming from surveillance research are most often derived from the magnitude of various factors and characteristics observed from their causal proof. Lewis et al.'s [63] study of injuries by needles and other sharp instruments in gynecologic and obstetric operations illustrates this kind of derivation. Medication errors are another topic studied often through some kind of surveillance.[64]

Cross sectional surveys,[65] *population-based surveys* based on medical records,[66] and other data sources or *longitudinal follow-ups without controls* serve principally as hypotheses generators of possible causal relationships rather than causal proofs themselves. Prospective *cohort studies*, with or without control groups including analyses between cohorts and within cohorts, provide a better idea of the possible importance of causal associations. For example, Makary et al.'s study of needle-stick injuries among surgeons in training[67] provides odds ratios indicating the importance of no involvement with a high-risk patient, no knowledge of injury caused by another person, and no occurrence in the operating room as variables most strongly associated with the nonreporting of the most recent needle-stick injury.

Case control studies are feasible based on all their advantages and limitations. In an already historical example, McCarroll and Haddon[68] use this methodology when studying causes (controllable "risk factors" and noncontrollable "risk markers") in drivers causing fatal automobile accidents. Alcohol was found to be an important risk factor, and being male appeared to be a risk marker. Gawande et al.'s study of risk factors for retained instruments and sponges after surgery[69] suggests the following as most important potential causal factors for instrument retention operation performed on an emergency basis: unexpected change in operation, more than one surgical team involved, and change in nursing staff during procedure. Counts of sponges and instruments appeared to be a potential preventive factor, and being female appeared to be a preventive marker.

Perneger[2] was right in calling for more epidemiological research of this kind wherever enough high-quality information or satisfactory numbers of patients and events allow. In other situations, better case reporting and case series must be considered whatever the limitations of causal proof may be.

Conclusions

As reported by Stelfax et al.,[70] the Institute of Medicine's *To Err Is Human* report was followed by an increase of various types of occurrence and etiology research such as qualitative studies, case and case series reports, correlation studies, cross sectional research, case control and cohort studies, interventions assessments, systematic reviews, and decision analyses. Results of such trends should prove promising for our better understanding of the medical error problem.

In light of current experience, what should novices in lathology searching for roots of error and harm keep in mind in their future initiatives? What should we expect in etiological research? What is expected from us on the basis of current factual and methodological experience in cause-effect relationships of interest? Without a doubt, the answer is a set of topics and variables of interest both in the human and system error domains, whenever a causal reasoning method is under consideration.

Depending on the research question, any of the following may play the role of a dependent or independent variable in observational analytical research, in clinical trials and health programs and other interventions based on rules of experimentation. The spectrum of variables of interest is wider in lathology than in many other domains of etiological research with respect to their nature, mode of study and applicability.

1. **Human (individual) error** as committed by decision makers, operators, executors or evaluators such as physicians, nurses, pharmacists, nutritionists, physiotherapists, health technology specialists:
   - Necessary clinical and fundamental knowledge, attitudes, skills for lathology problem solving
   - Physical (fatigue, stress), mental (empathy, affection, etc.), professional and other social (integration, beliefs, faith, values) disposition and states
   - Obtaining, evaluating and using the best available evidence
   - Missing evidence
   - Mastery and deficiencies in elements and mechanisms of reasoning, logic, argumentation, and critical thinking
   - Understanding of the above by health professionals as "manufacturers of health"
   - Decision making
   - Execution of tasks
   - Evaluation of executed tasks
   - Generalization, specific uses and applications of data and information stemming from the human error study
2. **System error**, encompassing:
   - Individuals and their errors involved in all the above-mentioned activities
   - Communication and its tools
   - Machines, instruments and other work tools and their evolving technology
   - Information technology and both its hardware and software
   - "Material on which we are working" i.e. recipients of our care such as patients with principal and collateral (associated co-morbidity) health problems and interventions and co-interventions of interest, healthy individuals with a potential target problem
   - Supporting human (social) structures such as administration, management, various levels of government (local, national, international) functioning and rules
   - Supporting physical (material) structures of the working environment
   - General culture, beliefs and faith providers (clergy and other executors and representatives) affecting to a varying degree decisions, acceptance and execution of lathology-related activities

– Webs of causes and webs of consequences constituted from all the above-mentioned elements.

Such an array of considerations is much wider and diverse than in many cancer or heart disease etiology studies or trials. Our knowledge and understanding of this understandable complexity and our experience with it are growing. The challenge may appear overwhelming at times, but can it be otherwise?

Whatever causal research demonstrates, possible or not, without knowing causes of error and its consequences incidence, how effective can primary prevention (control of error incidence) be? How can we conceive (if relevant) some kind of "duration" of error (through the duration of its consequences) and the "prevalence" of error (and its consequences) to better understand likelihood for secondary prevention (control of error and its consequences' prevalence by controlling the duration of error and its consequences), or tertiary prevention focusing on the minimization of the severity of error and its consequences cases?

The *a priori causal attribution model* [71,72] remains a powerful source of hypotheses in the hands of any epidemiological mind. As a necessary addition, the *epidemiological model of causal proof* expands our understanding of medical error from which error control should benefit wherever such epidemiological analyses are feasible, especially in the case of "new" and poorly understood origins of medical error. Our understanding of *necessary* and *sufficient causes* of medical error is still very limited. The distinction between *causal factors* and *contributing factors* remains imperfect. Let us do more, better, and more often to improve and expand etiological research in the medical error domain.

As far as medical error and harm themselves are concerned, the search for their causes is a multifaceted problem. The variable magnitudes and characteristics of their occurrence; their heterogeneous biological, technical, communicational, or intellectual nature; complex webs of causes and consequences; and multiple medical and non-medical means of control all contribute to and generate a methodologically difficult process that make this domain challenging both technically and practically. Our strategy for dealing with the error and harm problem will continue to evolve from one domain to another just like root-cause analysis has moved, for example, from management [73] into medicine and other health sciences [74]. This will require flexibility, adaptability, pragmatism, and focus on our part. Nonetheless, a more fundamental and clinical epidemiology, evidence-based approach, including critical thinking and decision-making, will not hurt.

References

1. Chang A, Schyve PM, Croteau RJ, O'Leary D, Loeb JM. The JCAHO patient safety event taxonomy: a standardized terminology and classification schema for near misses and adverse events. *Int J Qual Health Care,* 2005;**17**(2):95–105.
2. Perneger TV. Investigating safety incidents: more epidemiology please. *Int J Qual Health Saf,* 2005;**17**(1):1–2.
3. Jenicek M. *Foundations of Evidence-Based Medicine.* Boca Raton, FL: Parthenon Publishing Group Inc. (CRC Press), 2003.
4. International Epidemiological Association. *A Dictionary of Epidemiology.* Fifth Edition. Edited by M Porta, S Greenland, JM. Oxford: Oxford University Press, 2008.
5. U.S. Department of Veterans Affairs, National Center for Patient Safety (NCPS). *Glossary of Patient Safety Terms.* Available at: http://www.va.gov/NCPS/glossary. html, retrieved November 11, 2008.
6. Alemi F. Probabilistic risk analysis is practical. *Q Manage Health Care,* 2007;**16**(4):300–10.
7. Alemi F. for the Department of Health Administration and Policy, George Mason University Course on Risk Analysis in Health Care, HAP 735. Available at: http://gunston.gmu.edu/healthscience/RiskAanalysis/default.htm, retrieved June 22, 2009. See Session 6: *Probability of Rare Events.* Available at: http://gunston. gmu.edu/healthscience/Risk/Analysis/ProbabilityRareEvent.asp, retrieved June 22, 2009.
8. Altman DG, Bland M. Time to event (survival) data. *BMJ,* 1998;**317**(Aug 15):468–9.
9. Aronson JK, Hauben M. Anecdotes that provide definitive evidence. *BMJ,* 2006;**333**(Dec 16):1267–9.
10. Kiene H, von Schön-Angerer T. Single-case causality assessment as a basis for clinical judgment. *Altern Ther,* 1998;**4**(Jan):41–7.
11. Jenicek M. Clinical case reports and case series research in evaluating surgery. Part II. The content and form: Uses of single clinical case reports and case series research in surgical specialties. *Med Sci Monit,* 2008;**14**(10): RA149–RA162.
12. Bagian JP, Gosbee J, Lee CZ, Williams L, McKnight SD, Mannos DM. The Veterans Affairs root cause analysis system in action. *J Qual Improv,* 2002;**28**(10):531–45.
13. Department of Veterans Affairs, Veterans Health Administration. *VHA National Patient Safety Improvement Handbook.* VHA Handbook 1050.01, Transmitted May 23, 2008. Available at: http://www.va.gov/ncps/safetytopics.html, retrieved May 9, 2009.
14. Wikipedia, the Free Encyclopedia. Root cause analysis. Available at: http:// en.wikipedial.org/wiki/Root_cause_analysis, retrieved February 25, 2009.
15. Agency for Healthcare Research and Quality (AHRQ;USA.gov). *Glossary.* Available at: http://www.webm.ahrq.gov3/glossary.aspx, retrieved January 3, 2009.
16. Runciman W, Hibbert P, Thomson R, van der Schaaf T, Sherman H, Lewalle P. Towards an international classification for patient safety; key concept and terms. *Int J Qual Health Care,* 2009;**21**(1):18–26.

17. Croskerry P, Cosby KS, Schenkel SM, Wears RL. *Patient Safety in Emergency Medicine.* Philadelphia: Wolters Kluwer | Lippincott Williams & Wilkins, 2009.
18. Jenicek M, Hitchcock DL. *Evidence-Based Practice. Logic and Critical Thinking in Medicine.* Chicago: American Medical Association Press, 2005.
19. Wikipedia, the Free Encyclopedia. *Root Cause.* Available at: http://en.wikipedia.org/wiki/Root_cause, retrieved November 10, 2008.
20. U.S. Department of Veterans Affairs, National Center for Patient Safety. *Root Cause Analysis (RCA).* Available at: http://www.va.gov/NCPS/rca.html, retrieved January 20, 2009.
21. Canadian Patient Safety Institute: Hoffman C, Beard P, Greenall J, White J. *Canadian Root Cause Analysis Framework. A tool for identifying and addressing the root causes of critical incidents in healthcare.* Edmonton: CPSI/IMSP/Saskatchewan Health, 2008. Available at: http://patientsafetyinstitute.ca/uploadedFiles/resources/RCA_I, retrieved January 20, 2009.
22. Institute for Safe Medication Practices Canada analyzing team. *Fluorouracil Incident Root Cause Analysis.* Formatted for Web Posting May 22, 2007. Available at: http://www.cancerboard.ab.ca/NR/rdonlyres/2FB61BC4-70CA-4E58-BDE-IE54797BA47D/0/Fluorouraciincident/May2007.pdf, retrieved January 20, 2009. See also http://patientsafetyinstitute.ca/uploadedFiles/resources/RCA_I .
23. Flanders SA, Saint S. *Getting to the roots of the matter!* Spotlight case in emergency medicine, June 2005. Agency for Healthcare research and Quality (AHRQ) web M&M, Morbidity and Mortality Rounds on the Web. Available at: http://www.webmm.ahrq.gov/printview.aspx?caseID=98, retrieved July 19, 2009.
24. *Canadian Root Cause Analysis Framework. A tool for identifying and addressing the root causes of critical incidents in healthcare.* Edmonton: Canadian Patient Safety Institute, 2006.
25. Vincent CA, Adams S. Approaches to understanding success and failure. Chapter 8, pp. 49–84 in: Croskerry P, Cosby KS, Schenkel SM, Wears RL. *Patient Safety in Emergency Medicine.* Philadelphia: Wolters Kluwer | Lippincott Williams & Wilkins, (Health), 2009.
26. Berwick DM. Errors today and errors tomorrow. *N Engl J Med,* 2003;**348**(June 19):2570–2.
27. McDowell I. From risk factors to explanation in public health. *J Public Health,* 2008;**30**(3):219–23.
28. Percarpio KB, Watts VB, Weeks WB. The effectiveness of root cause analysis: What does the literature tell us? *Joint Comm Accred Healthcare Org,* 2008;**34**(7):391–8.
29. Wu AW, Lipshutz AKM, Pronovost PJ. Effectiveness and efficiency of root cause analysis in medicine. *JAMA,* 2008;**299**(6):685–7.
30. Stecker MS. Root cause analysis. *J Vasc Interv Radiol,* 2007;**18**(1):5–8.
31. Murphy JF. Root cause analysis of medical errors. *Ir Med J,* 2008;**101**(2):36.
32. Middleton S, Chapman B, Griffiths R, Chester R. Reviewing recommendations of root analyses. *Austr Health Rev,* 2007;**31**(2):288–95.
33. Coombes ID. Why do interns make prescribing errors? A qualitative study. *MJA,* 2008;**188**(2):89–94.
34. U.S. Department of Veteran Affairs, National Center for Patient Safety. *Root Cause Analysis.* Available at: http://www.va.gov/NCPS/rca.html, retrieved January 20, 2009.

35. U.S. Department of Veterans Affairs, National Center for Patient Safety. *Using the Five Rules of Causation.* Available at: http://www.va.gov/NCPS/CogAids/Triage/index.html?, retrieved January 20, 2009.

36. U.S. Department of Health & Human Services, AHRQ—Agency for Healthcare Research and Quality. Wald H, Shojania KG. *Chapter 5. Root Cause Analysis.* Available at: http://www.ahrq.gov/clinic/ptsafety/chap5.htm, retrieved November 9, 2008.

37. Shojania KG, Wald H, Gross R. Understanding medical error and improving patient safety in the in patient setting. *Med Clin N Am*, 202;**86**:847–67.

38. Williams PM. Techniques for root cause analysis. *BUMC (Baylor Univ Med Center Proceedings)*, 2001;**14**(2):154–7.

39. Weber GD. Discovering causal relations by experimentation: causal trees. MAICS-97 Proceedings, 1997, 91–98. Available at: http://www.aaai.org/Papers/MAICS/1997/MAICS97-017.pdf, retrieved June 19, 2009.

40. Kumar S, Steinebach M. Eliminating US hospital medical errors. *Int J Health Care Qual Ass,* 2008;**21**(5):444–71.

41. Kostopoulou O. From cognition to the system: developing an multilevel taxonomy of patient safety in general practice. *Ergonomics,* 2006;**49**(5-6):486–502.

42. Thomas M, Mckway-Jones K. Incidence and causes of critical incidents in emergency departments: a comparison and root cause analysis. *Emerg Med J,* 2008;**25**:346–50.

43. Wilson W. *Causal Factor Tree Analysis.* Available at: http://www.bill-wilson.net/b56.html, retrieved June 19, 2009.

44. Marx DA, Slonim AD. Assessing patient safety risk before the injury occurs: An introduction to sociotechnical probabilistic risk modelling in health care. *Qual Saf Health Care,* 2003;**12**(Suppl II):ii33–ii38.

45. Alemi F. Tutorial on discrete hazard functions. *Manage Health Care,* 2007;**16**(4):310–20.

46. Hovor C, O'Donnell LT. Probabilistic risk analysis of medication error. *Q Manage Health Care,* 2007;**16**(4):349–53.

47. George Mason University. *HAP 525: Risk Analysis in Healthcare.* Available at: http://gunston.gmu.ed/healthscience/riskanalysis/IntroductionCourseRiskAnalysis.asp, retrieved June 19, 2009.

48. U.S. NRC. *Probabilistic risk analysis.* Available at: http://www.nrc.gov/reading-rm/basic-ref/glossary/probabilistic-risk-analysis.html, retrieved June 19, 2009.

49. Wikipedia, the Free Encyclopedia. *Probabilistic risk assessment.* Available at: http://en.wikipedia.org/wiki/Probabilistic_risk_assesment, retrieved June 19, 2009.

50. Bowie P, Pope L, Lough M. A review of the current evidence base for significant event analysis. *J Eval Clin Pract,* 2008;**14**:520–36.

51. Royal College of General Practitioners. *2. Significant event analysis.* Available at: http://www.org.uk/practising_as_a_gp/distance_learning/egp2_update/learning_tool...,* retrieved June 19, 2009.

52. Anon. *Significant Event Analysis (SEA).* Available at: http://www.pennine-gp-training.co.uk/Significant-Event-Analysis.doc, retrieved June 19, 2009.

53. *Significant Event Analysis.* Available at: http://www.pdptoolkit.co.uk/Files/Guide%20to5220PDP/content/significant_event...,* retrieved June 19, 2009.

54. NHS Education for Scotland. *Significant Event Analysis. Standard Report Format.* Available at: www.nimdta.gov.uk/.../significant_event_analysis_standard_report_format.doc, retrieved June 19, 2009.

55. Vincent C, Taylor-Adams S, Chapman JE, Hewett D, Prior S, Strange P, et al. How to investigate and analyze clinical incidents: Clinical Risk Unit and Association of Litigation and Risk Management Protocol. *BMJ,* 2000;**320**(18 March): 777–81.

56. Vincent CA. Analysis of clinical incidents: a window on the system not a search for root causes. *Qual Saf Health Care,* 2004;**13**:242–3.

57. Wikipedia, the Free Encyclopedia. *Thought experiment.* Available at: http://en.wikipedia.org.wiki/Thought_experiment, retrieved November 20, 2008.

58. Sokol D. *What if …* Available at: http://newsvote.bbc.co.uk/mpapps/pagetools/print/news.bbc.co.uk/2/hi/uk_news/magazin..., retrieved January 26, 2009.

59. Stanford Encyclopedia of Philosophy. Brown JR. *Thought Experiments. March 25, 2007 Revision.* Available at: http://plato.stanford.edu/entries/thought-experiment/, retrieved January 26, 2009.

60. Leitch RA, Moses GR, Magee H. Simulation and the future of military medicine. *Military Medicine,* 2002;**167**(4):350–4.

61. Gardner R, Raemer DB. Simulation in obstetrics and gynecology. *Obstet Gynecol Clin N Am,* 2008;**35**:97–127.

62. Waterston L. Monitoring sharp injuries: EPINet surveillance results. *Nurs Stand,* 2004;**19**(3):33–8.

63. Lewis FR Jr., Short LJ, Howard RJ, Jacobs AJ, Roche N. Epidemiology of injuries by needles and other sharp instruments. Minimizing sharp injuries in gynaecologic and obstetric operations. *Surg Clin North Am,* **75**(6):1105–1121.

64. Rask K, Hawley J, Davis A, Naylor D. Impact of statewide reporting system on medication error reduction. *J Patient Safety,* 2006;**2**(3):116–23.

65. Sexton MJ, Thomas EJ, Helmreich RI. Error, stress, and teamwork in medicine and aviation: cross sectional surveys. *BMJ,* 2000(March 18);**320**:745–9.

66. Thomas EJ, Brennan TA. Incidence and types of preventable adverse events in elderly patients: population based review of medical records. *BMJ,* 2000 (March 18);**320**:741–4.

67. Makary MA, Al-Attar A, Holzmueller CG, Sexton JB, Syin D, Gilson MM, et al. Needlestick injuries among surgeons in training. *N Engl J Med,* 2007;**356**(26):2693–9.

68. McCarroll JR, Haddon W Jr. A controlled study of fatal automobile accidents in New York City. *J Chron Dis,* 1962;**15**:811–26.

69. Gawande AA, Studdert DM, Orav EJ, Brennan TA, Zinner MJ. Risk factors for retained instruments and sponges after surgery. *N Engl J Med,* 2003;**248**(3):229–35.

70. Stelfax HT, Palmisani S, Scurlock C, Orav EJ, Bates DW. The *"To Err Is Human"* report and the patient safety literature. *Qual Saf Health Care,* 2006;**15**:174–8. Doi: 10.1136/qshc.2006.017947.

71. Reason J. The contribution of latent human failures to the breakdown of complex systems. *Phil Trans Roy Soc London,* 1990;**327**(1241) Series B:475–484.

72. Rasmussen J. Human error and the problem of causality in analysis of accidents. *Phil Trans Roy Soc Lond,* 1990;**327**(1241) Series B:449–62.
73. Wilson PF, Dell LD, Anderson GF. *Root Cause Analysis. A Tool for Total Quality Management.* Milwaukee, Wisc: ASQC Quality Press, 1993.
74. Joint Commission Resources (Joint Commission on Accreditation of Healthcare Organizations—Joint Commission). *Root Cause Analysis in Health Care. Tools and Techniques.* Third Edition, Oakbrook Terrace, Ill: Joint Commission Resources, Inc. (JCR), 2005.

Chapter 6

Flaws in Operator Reasoning and Decision Making Underlying Medical Error and Harm

Executive Summary

Two major trends appear in today's approach to the medical error problem. Either healthcare is seen as a system that produces medical error, or it is thought that the final operator and decision maker provides faulty explanations and decisions that lead to patient harm. Both are intertwined facets in which individual or operator error cannot be left as a black box in the system error. This chapter is about how medical error should be understood on an individual (operator) basis before it is integrated into system analysis and interpretation. As an introspection or review with peers, it should precede any further scrutiny by health administrations as well as legal system assessments and assessments in the broader context of community life, expectations, values, and willingness to improve clinical and community care and its quality and results.

Medical care is a process that includes the patient's entry under care, the physician's physical clinical and paraclinical evaluation and diagnosis, the decision to treat and how to treat, the action of medical care itself, the evaluation of its impact or effect, and an outcome assessment in the framework of prognosis and further care. Errors may occur as a result of sensory-perception-communication failure, medical thinking, reasoning and decision-making flawed processes, and wrong or right actions executed, evaluated, and followed by foreseeable predictions in terms of prognosis and outcome follow-up.

Evidence build-up, reasoning about the case and diagnosis, decision making regarding what to do, sensory-motor execution of medical acts, obtaining results of care, evaluation of its results, prognosis and further risk assessment and necessary follow-up, surveillance, and forecasting all rely on their clinical-epidemiological validity and quality of argumentation that support each of such medical care steps. Evaluation of errors and the entire quality of clinical and community care depends then on the evaluation of evidence, its uses, and argumentation supporting step-by-step decisions in the clinical and community care process.

Paths in the care process structure, analysis, and their endpoints differ from paths in research, however complementary they may be. This chapter is about what operator understanding and decision making might generate as errors, what to do with them, and how to integrate them into system evaluation later on, as already described in the preceding chapters. Chapter 7 is about interventions in the medical error and harm domain and their evaluation.

Thoughts to Think About

Only those who do nothing at all make no mistakes … but that would be a mistake.

—Anonymous, n.d.

If you don't make mistakes, you're not working hard enough on problems. And that's a big mistake.

—Author unknown

Things could be worse. Suppose your errors were counted and published every day, like those of a baseball player.

—Author unknown

Error is always in haste.

—Thomas Fuller, 1723

A man of genius makes no mistakes. His errors are volitional and are the portals of discovery.

—James Joyce, *Ulysses*, 1922

Only, there are not many geniuses among us! Strong people make as many mistakes as weak people. The difference is that strong people admit their mistakes, laugh at them, learn from them. That's how they become strong.

—Richard Needham, 1999

Not at the patient's expense, please! Don't argue for other people's weaknesses. Don't argue for your own. When you make a mistake, admit it, correct it and learn from it immediately.

—Stephen R. Covey, undated, b. 1932

Don't ever make the same mistake twice, unless it pays.

—Mae West, 1892–1980

Never say, "oops." Always say, "Ah, interesting."

—Author unknown

Recently, I was asked if I was going to fire an employee who made a mistake that cost $600,000. No, I replied, I just spent $600,000 training him.

—Thomas J. Watson Jr., 1914–1993

Would you do the same thing with a surgeon, emergency physician or intensive care nurse who lost a patient due to error? After a long, arduous and expensive education, doctors are expected to get it right. But they are fallible human beings like the rest of us. Mistakes are stigmatized rather than being seen as chances for learning … yet doctors are given very little training in understanding, anticipating, detecting and recovering from errors.

—James Reason, 2009

Wanting to learn from our errors, aren't we often hesitant, whatever the reason may be, to have them assessed by retrospection based on our strong or poor reasoning and decision making and on their possible deficiencies to minimize harm and maximize patient safety? Before sharing our error analysis with others, what prevents us from performing our own error analyses in the privacy of our own introspection? Nothing. Let us begin then with an analysis and understanding of the path of our own actions.

Can and should medical error be attributed to multiple failures of the system or to human failure at the end of some causal chain? To both, we believe. Medical error is a problem not only of faulty logic or cognition but also of medicine itself, ergonomics, system components, and functioning failures and flaws.

Is it enough to prove that an error occurred? No, it's not. Proof of who or what was (or was not) responsible for error is equally important to correct it.

Introductory Comments

This chapter shows and offers the reader guidelines to detect faults in the reasoning and decision making of health professionals and to identify where corrections should be made. Such health professional's errors detections and identifications should help us to understand better corrective actions and interventions as well as their evaluation outlined in Chapter 7.

However complex the chain or web of latent (remote) or active (proximate, immediately preceding) errors leading to incidents or accidents with all their consequences may be, there is always an operator at the end who introduces (or not) all such necessary elements into his or her reasoning, understanding, and decision making with regard to what to do in a particular case of medical care. In the framework of our quite understandable no-blame culture, a lot of attention has been paid so far to system components such as technologies, communication, and environmental, psychological, and physiological factors in care providers. Much less has been said about where sites of error in chains of our reasoning and decision making may be, as applied to various components of patient care including risk assessment, diagnosis, treatment, prognosis, and prevention.

A recent study of errors in surgery by Fabri and Zayas-Castro[1] analyzed surgical complications in 332 patients over a 12-month period of time. It attributed 63.5% of complications to surgical technique, 29.6% to errors in judgment, 29.3% to inattention to detail, and 22.7% to incomplete understanding. A total of 58% of errors were slips, and 20% were mistakes. Human factor causes of error prevailed, and organizational/system errors or breaks in communication were less frequent; system errors and communication errors at 2% each were noted. More of such studies are needed in other medical specialties.

In the *Quality in Australian Health Care Study,* iatrogenic (surgery and beyond) patient injuries termed *adverse events* were associated with 16.6% of hospital admissions. The major causes determined were (1) complications of, or failure in, the technical performance of an indicated procedure or operation

(34.6%) and (2) failure to synthesize, decide, or act on available information (15.8%). Some other causes of interest were acting on insufficient information (1.8%), slips and lapses due to "absentmindedness" in activities in which the operator is skilled (1.6%), and lack of knowledge (1.1%). Multiple procedural failures were also mentioned.[2] In another Australian pilot study of medical errors, a 21% occurrence of knowledge and skills errors was identified.[3]

Let us examine more closely in this chapter medical error committed by operator flaws in reasoning and decision making.

Medical or nursing students are more accustomed to having their decisions scrutinized and accept it as part of the learning process and as evaluation of their knowledge, attitudes, or skills. Accomplished health professionals are often hesitant to share information because it may lead to third-party analysis for psychological, legal, or monetary and otherwise punitive reasons. In addition, their lack of willingness to share is encouraged by hospital, insurer, and attorney suggestions to avoid words like *error, harm, negligence, fault,* or *mistake,*[4] whose meanings may vary and which may trigger litigation. In all cases, readers who wish to correct and improve their actions should perform such evaluations for themselves first, looking at how their reasoning, decision making, and behavior fit into the man–machine system. Then, that information should be shared with someone else. In the health domain, we are faced with an "operator-health technology-patient system" as a tripartite functioning entity.

As much as we stress the need for knowing "external" causes of medical errors and their consequences, let us see what happens "in the cranial cavity" of the healthcare provider as an "operator" at the decisive endpoint before action.

Errors of action stem from two possible sources:

1. They are of a sensory-motor skill nature as seen through ergonomics (skill based in Reason's terminology).
2. They result from faulty reasoning, critical thinking, and argumentation (rule and knowledge based).

This chapter is about the latter. Thanks to critical thinkers like philosophers, psychologists, and epidemiologists, the accumulation of valuable, usable, and operational methodology acquired in this domain throughout past generations allows us today to treat this question as more than a black box.

Note about Medical Error and Medical Harm

As we have already mentioned, medical error and medical harm are sometimes homozygous, sometimes heterozygous, and sometimes heredity-unrelated twins. Less symbolically speaking, the link between them produces and invokes the attention and involvement of multiple specialties when dealing with their challenges.

Besides medicine itself, the study and management of the medical error will require significant involvement from logic, critical thinking, psychology, operational research, and computer sciences. Management of medical harm will rely on the essential mastery of clinical specialties in which harm occurs (i.e., clinical epidemiology, biostatistics, and evidence-based medicine). In lathology, both are brought together in different proportions and balanced from and between the domains of error and harm.

System Error versus Individual Human Error

The current ongoing debate regarding whether medical error is a system error or an individual human error at its effective or "sharp end" is still open. There are different views pertaining to whether we should consider medical error as a system/multiparty/multifactorial error or as an individual/active/human/operator error. Ultimate decisions in clinical and community care always belong to the physician and any other health professional responsible for practice, care, and research.

Moreover, no one has precisely established the borders between latent and active errors. Is active error only the error committed by an operator at the end of the medical care process, or is it anything that happened since the admission of the patient or at another specific moment in patient and community care? For now, let us define it for each individual case of error study.

Errors caused by operators (i.e., actors at the "sharp end" of the process) may be of numerous types:

- *Sensory–perception–communication failure* stemming from the observed and otherwise perceived problem (e.g., literature, past experience, observation of the patient or the community). Misreading labels in clinical pharmacology or patient charts before surgery, omission of medical orders, fatigue, or stress are just some examples. This is a noticing challenge.
- *Medical thinking–reasoning–decision-making flawed processes* in which physician activities "are just not logical." Examples include reasoning and critical thinking supporting some kind of conclusion such as, "This patient has or doesn't have this disease," "He or she is at risk of some future problem and that's how he or she should be treated," or "That's the prognosis for the case based on poor evidence, poorly linked and irrelevant for the conclusion." This is an interpretation challenge.

- *Decision challenges regarding how to deal with flaws.* Decisions pertaining to what to do are not based on an explicit decision-making process, be it decision analysis, heuristics, algorithms, consensus, hindsight thinking, or something else.
- *The action undertaken is wrong, or the action is not properly executed.* The operator's sensory and motor skills fail in the execution of preceding decisions and orders for action: not administering a prescribed medication from orders, failing a surgical intervention, making unrealistic prognoses (not yet litigated at courts of law). It is an execution break-up whatever sensory motor *fausse route* is behind the error and its consequences of interest. This is a challenge of how and what to do with what was decided.
- The *intervention is inadequately evaluated* for its soundness, structure, process (execution), and impact. This is a challenge of knowing what was achieved by all this.
- The *prognosis is made without often necessary distinctions* from patient history, between risk factors and prognostic factors, and comorbidity.

Error may occur due to the failure in any of the aforementioned stages of the medical care process. Therefore, to correct it, all stages should be scrutinized.

This chapter is about points to ponder in the medical thinking–reasoning–decision-making processes, the potential and real errors to which we must always pay attention if we want to fully understand and "fix" causes of medical error. Usually, individual reasoning and decision-making processes are treated across current literature and experience as some kind of black box. They should not be. There is always someone behind interpretations and actions to be decided and executed. The current increasingly rich critical thinking and decision-making experience and its increasingly structured methodology as well as progressing clinical experience and applications in this domain leads us to think in this manner.

The question is worth repeating: is it enough to blame the system only for a medical error? Definitely not! However psychologically, professionally, or socially it might be resented, both system and individual/human/active error must be looked upon, understood, corrected, and prevented in the future.

Reminder regarding Some Fundamental Considerations

Let us start with a practical example. An internist examines a patient suffering from coronary heart disease. He or she is faced with the dilemma of pursuing the conservative (medical) treatment of coronary pathology and dysfunction or to refer the patient to invasive cardiologists for angioplasty or to surgeons for coronary bypass surgery. Where might errors occur, and how should they be prevented? Later on, a catheter for angioplasty malfunctions, the patient does not need angioplasty after all, and the patient's prognosis worsens. Why is this the case?

Two visions of the cognitive and decision-making processes must be retained, namely, that of an argument and argumentation process based on evidences and that of a step-by-step review of our diagnostic, therapeutic, or prognostic processes both for short- and long-term care:

1. A decision like "this patient must be transferred to surgery" can be considered a conclusion or claim (as philosophers used to say) of the *argumentation process* using evidence of all kinds.[5,6] Errors may occur due not only to uses of poor evidence or failures of using good evidence from one step of argumentation to another but also to the linking together of various components on the way to conclusions and recommendations. The argumentation process and model applies practically to all steps and stages of clinical practice and health and community care.

2. The additional, more specific *cognition pathways in various clinical practices and components of care* from their original triggers to their final understanding and ensuing decisions to be considered must be recognized in the development of any necessary steps in risk assessment (as seen partly in the preceding chapter), diagnostic process, treatment plan work-up, or prognosis with its ensuing short- and long-term actions in care. Faults in any of these steps may be causes of numerous incidents and their consequences.

A step-by-step evaluation of both argumentation and cognition processes must contribute to improvements in the medical error domain. All these processes are used at each of the following steps in medical care:

- *Evidence build-up*, acquired knowledge, and basic understanding structured or unstructured (history, literature, physical and paraclinical examination, past experience).
- *Reasoning* about the case, problem or situation, structured acquisition of *new knowledge* (differential diagnosis, final diagnosis, comorbidity assessment).
- *Decision making* (medical, surgical, psychiatric, and other orders and plans for care).
- *Sensory-motor execution of medical acts* (e.g., operation, invasive diagnostic procedures, physiotherapy, parenteral applications of drugs, new technologies devises implantation).
- *Getting results* (positive and adverse).

- *Evaluation (of all the preceding steps, one by one* and the entire process; successes, errors, failures).
- *Prognosis and further risk assessment* (morbidity, comorbidity, further outcomes).
- *Follow-up, surveillance, forecasting* (risk factors and markers, prognostic factors and markers, possible outcomes, errors and failures).

Correct or incorrect argumentation, reasoning, critical thinking, and decision making underlie such processes.

Flawed Argumentation and Reasoning as Sites and Generators of Error and Harm: Argumentation and Human Error and Harm Analysis from a Logical Perspective

We do not always realize that modern argumentation and argument are behind most of our evidence and critical-based thinking, clinical epidemiology grounded in reasoning, and decision making, as we have already discussed elsewhere.[5-7] For greater simplicity, let us summarize here again some basic terms, principles, rules, and guides scattered throughout previous chapters.

Diagnosis and differential diagnosis, treatment choices, or predictions in prognosis can all be viewed as the products of arguments. Why is this the case? In our reasoning, we may proceed in several ways:

1. In the simplest situations of daily practice, we proceed from some reason to some conclusion: "This patient complains of a sore throat [reason, proposition, premise], so [indicator, connector] let us see if he has a streptococcal infection [conclusion about what to do, claim, recommendation, orders]."
2. We may also recall Aristotelian thinking as illustrated in a categorical syllogism: "All sore throats require attention if they are not caused by a streptococcal infection [major premise]; my patient has a sore throat [minor premise] so I must be sure that he does not have this kind of bacterial infection before reassuring him that he does not need antibiotics [conclusion]."
3. Another method of reasoning and decision making involves applying the modern way of argumentation proposed by Toulmin,[8,9] detailed elsewhere[5]: "I will prescribe you antibiotics for your bacterial throat infection [claim, conclusion of an argument] because looking at your red throat, patched tonsils, tender cervical nodes and fever, and positive results of our

rapid laboratory test [grounds] a 10-day treatment with penicillin may be the best choice in your case [warrant]. All our past experience and clinical studies show that we must do this to spare you from serious complications of such an infection [backing] so let us definitely proceed this way if you agree [qualifier]—unless you are allergic to penicillin [rebuttal], in which case we should choose another kind of treatment."

Do those manners of reasoning relate to our dealings with medical error? They definitely do. Theoretical foundations of argumentation are explained in more detail in our[5-7] and other[8,9] original writings.

Here is a reminder of some essential terms in the context of this chapter:

Argumentation falls into the domain of *reasoning*, that is, as a tool to form conclusions, judgments, or inferences from facts or premises.

An *argument* is a connected series of statements or reasons intended to establish a position leading to another statement as a conclusion.

An argument is a vehicle of our *medical logic* as a system of thought and reasoning that governs our understandings and decisions in clinical and community care, research, and communication.

Informal logic helps us deal with health problems by looking at arguments as they occur in the context of natural language used in everyday life.

Argumentation is a methodological employment or presentation of arguments. Correct argumentation with ourselves and with other interested parties is one of the important ways to deal with a given health problem, disease, medical care, and medical error as well.

In light of current evidence in the medicine paradigm, *argument-based medicine*[7,10] refers to the research and practice of medicine in which understanding and decisions in patient and population care are supported by and based on flawless arguments using the best research and practice evidence and experience as argument and argumentation building blocks in a structured, fallacy-free manner.

As already suggested, medical reasoning fits extremely well within the framework of the *modern argument* as developed by Toulmin.[8,9] Figure 6.1 illustrates both its theoretical components and an application.

In simpler terms:

■ We need some *grounds* for argumentation. Clinical and paraclinical data and information particular to the case serve this purpose.

■ We submit grounds in light of a *warrant*, that is, some kind of general rule, accepted understanding, and evidence. Plausibility is in focus.

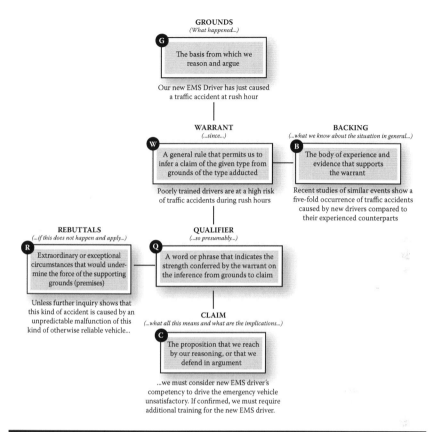

Figure 6.1 Components and structure of Stephen Toulmin's[8,9] modern argument with application to the error domain.

■ Whatever we conclude on the basis of grounds and warrant is evaluated in a confirmation, line, or distinct pattern as seen through *backing*, that is, research findings, graded evidence, past practical clinical experience, "external evidence," "what literature tells us."

■ Putting all this together, we try to somehow quantify the certainty or probability that our ensuing claims or conclusions of the argument are correct in terms of a *qualifier*, often the hardest challenge of an argument.

■ Our *conclusions* or argument *claim* are then the result of synthesis of the previous argument's building blocks.

■ Argument conclusions are valid provided only that there are existing exclusionary circumstances or criteria that Toulmin termed *rebuttals*. Besides qualifiers, rebuttals are often sorely missing (and they should not be) in our conclusions or claim statements.

Therefore, an error occurs and mistakes happen if:

- Grounds are of poor quality, incomplete, or unrelated to the problem under study.
- Our understanding of the essence of the problem is not clear.
- The backing is of poor quality, incomplete, or unrelated again.
- We are unsure of our certainty regarding our conclusions and claims.
- We act as if exceptions would not apply to what we conclude about the problem.
- There is no meaningful link between each of the argument building blocks.

Ensuing incidents and accidents and their consequences (harm) may be seen then on the basis of:

- The formal quality of the argument.
- The quality and complexity of evidence that is fed into the argument.
- The inclusion or exclusion of other system errors, be they latent or active and that may prove necessary as parts of argument building blocks.

Operators such as an internist, clinical pharmacologist, or psychiatrist establishing a treatment plan; a surgeon choosing and using an invasive exploratory or operating instrument; an intubating emergency medicine physician; or a trauma-assisting pediatrician are all ultimate providers of correct or incorrect decisions, actions, and their potential harm, however worthy other system components might be of inclusion in our reflection about causes of medical error.

If we do not evaluate operator errors, how can we evaluate person and system errors? How can we correct and more fittingly improve the operator's work? This is what lies behind such questions as, "How did you arrive at your decisions and actions in the care of this patient?" Moreover, before attributing a medical error to some system, shouldn't we first prove that an individual is not responsible for an error and the harm to which it leads?

So far, the problem of faulty logic and critical thinking is only briefly acknowledged in current taxonomies of error proposals. For example, currently available taxonomies (see Chapter 3) summarily include knowledge, failure, and retrieval and usage of stored instructions but not more specific failures in reasoning and related decision making.

Mistakes and Errors in Medical Lathology

Lexically speaking, mistakes and errors are synonyms. More refined distinctions have, nevertheless, been proposed for the human error domain.[11] In the spirit of and in line with Chapter 2, let us remember the following:

- A *mistake* is an error in action, calculation, opinion, or judgment caused by, for example, poor reasoning, carelessness, and insufficient knowledge[12]; an incorrect judgment; an incorrect statement proceeding from faulty judgment; inadequate knowledge or inattention[13]; or a misconception or misunderstanding.[12]
- An *error* as synonym of mistake may be seen in our context as an act that through ignorance, deficiency, or accident departs from or fails to achieve what should be done. A mistake implies misconception or inadvertence and usually expresses less criticism than error.[14]
- *Lapses* are failures of memory.

In this framework, Reason[11] distinguishes between slips and mistakes as already outlined in Chapter 3. *Slips* are generally attentional and perceptual failures,[11] and *mistakes* are due to failures of mental processes.

Hence, errors as failures in reasoning, logic, or argumentation fall under the current category of mistakes. An erroneous diagnosis with its consequences is a mistake. Forgetting to execute clinical orders written in the patient's chart is a lapse according to current terminology and semantics. Misreading a decimal point on a medication dosage label is a slip according to current general human error considerations and terminology. Forgetting to administer prescribed medication is a lapse.

Understanding medical error as a result of faulty judgment, reasoning, and argumentation whose conclusions lead to patient harm is at the core of our preoccupation with human medical error attributable to individual health professionals.

Fallacies, Biases, and Cognitive Errors in Medical Lathology

Medical error as a mistake may be due to faulty reasoning and decision making. Such faults fall into the category of fallacies, biases, and cognitive errors. They may be defined and illustrated by an example related to the medical error domain:

- Fallacy: In more general terms, a *fallacy* is some mistake or flaw in reasoning or argument. In the broadest of terms, it is a violation of the norms of good reasoning, rules of critical discussion, dispute resolution, and adequate communication.[15] An example: A *cum hoc ergo propter hoc* fallacy means drawing conclusions about cause–effect relationships on the basis of loosely defined "associations" or "correlations" instead of on the basis of a more formal cause–effect proof. Anything that precedes medical error and its potential consequences (harm) should not automatically be considered a "cause" of harm until it conforms to criteria of causality.
- Bias: The meaning of the word *bias* in medicine is most strongly influenced by its meaning in biostatistics: a deviation from some real or reference

value. In a more general context, bias may be considered a nonrandom, systematic deviation from truth, or a well-defined "reality." Often, bias is a term used in an undistinguished way from cognitive error or cognitive bias. An example: drawing conclusions about a population on the basis of a sample that is biased or prejudiced in some manner. This is sometimes called a fallacy of *biased sample* or *underreporting of facts and variables* that are relevant for problem solving. Ensuing arcane explanations are biases in the aforementioned sense.

■ Cognitive error (cognitive bias): A *cognitive error* or *cognitive bias* (terms often used interchangeably across the literature) denotes a pattern of deviation in judgment that occurs in particular, within our area of interest, in medical and clinical situations or in medical research reasoning and conclusions. An example: *Hindsight bias* (loosely synonymous with outcome bias) is a cognitive error as well. It is a tendency for people with outcome knowledge to exaggerate the extent to which they would have predicted the event beforehand; it can have an adverse impact on retrospective investigations of events that cause harm.[16] *Outcome bias* refers to the influence of outcome knowledge on evaluations of decision quality, sometimes enabling mediocre processes to be judged as good and good processes as mediocre.[16]

Bias as it is used currently across the medical literature has multiple meanings including almost any flaw in reasoning and decision making, especially in medical research: research design, execution, and evaluation. It is increasingly discussed[15-22] because even "biased" research results and its uses may be detrimental to patient safety and health.

Susceptibility biases, protopathic bias, design bias, trial execution bias, overcharging bias, information bias, transfer bias, and performance bias are just a few examples that may compromise internal or external validity and serve as potential sources of harm for medical research results. An exhaustive list of biases as flaws in research is well beyond the scope of this text and must be sought in the previously quoted literature and paired with the more general list of cognitive biases found in Appendix A of this book and elsewhere.[23]

In addition to the bias challenge itself, across current experience and literature, extensive arrays of known fallacies (see Appendix B), biases, and cognitive errors often overlap.[7] Some may correctly perceive this kind of semantic and terminological disarray that still awaits correction and better clarification. Valuable collections of fallacies, biases, and cognitive errors with their definitions, examples, and ways to correct them increase in numbers and quality as reviewed in our preceding monograph,[7] together with a more detailed discussion of the fallacy/bias/cognitive error domain. We should be aware of as many important reasoning errors as possible in particular as they pertain to clinical and research situations to correct or avoid them. Current lists[7,23-35] are still not exhaustive and evolve in time. Any particular clinical or research situation may create some new ones to discover, define, understand, and avoid such as medical error.

The relevance of fallacies, biases, and cognitive errors may vary from one medical specialty to another. For example, Gunderman[36] points out attribution, availability, commission/omission, confirmation, framing, hindsight, regret, and satisfaction of search biases behind potential and real errors as particularly relevant for radiology. For surgery, Fabri and Zayas-Castro,[1] already quoted in the introduction to this chapter, mention the following as the most frequent errors: technique error, carelessness/inattention to detail, judgment error, and incomplete understanding of the problem. Susceptibility, design, information, transfer, performance, citation, optimism, and conflict of interest biases in surgery were discussed by Paradis.[22] What about in psychiatry?

Our understanding of medical error should be free from as many fallacies, biases, and cognitive errors as possible, an ideal that will not ever be reached but should be always be fully attempted given current specific conditions, the nature of the problem, and available means. "It can't be done" is not an answer.

Where and When Errors Occur: Cognitive Pathways as Sites of Error

Improving clinical practice by finding out only that our patient risk assessment, diagnosis, treatment plan, prognosis, or prevention decisions are wrong is indicative of a problem, but it does not help much to identify where in the process of practice the error occurred, what was responsible for it, and where there is room for enhancement. All such steps in medical practice may be correctly viewed as a cognitive process, a path from initial stimuli and ideas to observations, making sense of them and reaching some new knowledge as a basis for understanding or decisions concerning what to do. At what exact moment did the error occur? What part, stage, or step in clinical activities was or might be in the future a weak point to watch?

In this chapter, let us look at diagnosis, treatment, prognosis, and risk assessment relevant to and interesting from the point of view of understanding medical error better at the human level. In the next chapter, we will examine strategies of medical error prevention and control.

Physicians follow several steps and considerations when taking care of their patients:

- Learning about the patient's past and present through medical history. Gathering written and oral information about the past and present (chief complaint ant other complaints).
- Physical examination or mental status assessment, leading to and making diagnosis.
- Taking into account risks (future health problems) and health problems already present (diagnosis).
- Treatment choices, decisions, plan, and orders.
- Assessment of prognosis.
- Evaluation of the effectiveness, efficiency, and efficacy of the collection of previously mentioned steps in healthcare.
- Future steps in care to consider.

At **each** step in clinical work, conclusions are a sort of a claim (the endpoint in argumentation) or are based on what we have done about both positive or negative aspects and the impact of our care:

- Those data and information are relevant for further assessment of the patient.
- The patient has this disease, syndrome, and other health problems.
- Considering what I have heard, read, and seen in this patient, he or she may develop health problems that he or she does not have yet.
- The patient should receive the following medical, surgical, psychiatric, or other treatment and care.
- The outlook for the patient (prognosis) with or without treatment is good or bad.
- Our treatment plan and its execution were effective (did some good in habitual conditions), efficient (in line with money, resources and time invested), and efficacious (consistent, specious, and beneficial result under ideal conditions).

Errors may occur at any of these steps of research and practice. If we do not analyze, interpret, and know more intimately lathologic aspects of these steps, can we attribute our errors to the system etiology of error only? We believe not.

Reviewing Diagnoses: Searching for Errors in the Clinimetric Process

It is not enough to say that our diagnosis is wrong. Once we discover that our diagnosis is erroneous, we want to know (and must know) where exactly our error occurred.

Table 6.1 Tracking Errors, Their Sites, and Sources across the Steps of Clinical Work

Steps of Clinical Work	Possible Site and Type of Error
Patient's entry under healthcare	Recruitment, eligibility
Problem identification	Patient Complaints
Impression (diagnostic)	Hypothesis Working idea
Clinical work-up plan	Plan of investigation and examinations to order
Data collection (interview, clinical, and paraclinical examination)	Qualitative and quantitative • measurements • counting • categorization • classification
Analysis	Differential diagnosis (options) Horizontal assessment of evidence Soundness, structure, process, and impact of possible methods of care
Working (final) diagnosis	Diagnostic technique used (pattern recognition, exhaustive exploration, steepest ascent method, algorithms)
Decision Making	Vertical assessment of evidence (and of treatment) Decision analysis (including treatment options, probabilities, values, utilities chaining, and path tracing) Chagrin analysis Patient care algorithm work-up Patient preferences and values integration with his or her physician's preferences, values, and options

Continued

Table 6.1 Tracking Errors, Their Sites, and Sources across the Steps of Clinical Work (*Continued*)

Steps of Clinical Work	Possible Site and Type of Error
Decision	Conclusions about the best and other evidences
	Final diagnosis
	Treatment plan and orders (harmonizing physicians and patient's preferences, values, and options)
	Various other claims and critical thinking conclusions about diagnosis, treatment plan, outcomes, and prognosis
	Ensuing additional care given the prognosis
Action	Procedures implementation that include corrections according to occurring and progressing, unexpected and expected results
Evaluation	Effectiveness, efficacy, efficiency of care
	Expected and unexpected adverse effects
	Patient tolerance of procedures, satisfaction, possible future preferences
	Outcomes assessment and prognosis
	Assessment of the soundness, structure, process, and impact of the entire clinical care (in this patient and in other patients— overall practice)

At the beginning of the eighties, Alvan Feinstein[37–40] originally proposed the term *clinimetrics* for the measurement of clinical phenomena or "the domain concerned with indices, rating scales and other expressions that are used to describe or measure symptoms, physical signs, and other distinctly clinical phenomena in clinical medicine." The production of raw individual data from medical history, physical examination, paraclinical explorations, and other primary sources was then called *mensuration* in opposition to the next step, *measurement*—that is, processing raw data not only to give them (e.g., signs, symptoms) some dimension (e.g., severity) but also to label them qualitatively through classification, group formation, and validation as syndromes and other diagnostic entities indicative of further specific attention and care.

In this context, medical error not only means that some diagnostic instrument such as a biochemical test or fiber optic imaging scope broke down or did not work properly; the culprit may also be a faulty activity at any moment of the clinimetric process. Identifying such moments and activities caused by the operator, and not simply the machine, is necessary.

It is not enough either to relate a diagnostic error to poor internal validity of a diagnostic test relying on a specific substance or instrument in terms of sensitivity, specificity, predictive values, or reproducibility. Our sensory perceptions and their intellectual management count too, and all (or none) may be faulty. Figure 6.2 illustrates, based on a purely fictitious example, eight steps in the clinimetric process of moving from initial observations to diagnosis. At each of them, an error may occur, and harm ensues.

Step 1, trigger of care, may play a role in the subsequent diagnostic process itself. Steps 2–4 are steps in observation or mensuration. Steps 4–7 are those of explanation. Step 8 is the final diagnosis, the end of the path from the first perception of the patient following his or her complaint or motive for consultation to the final conclusion (claim) about the patient's state.

An incorrect diagnosis leading to incorrect treatment and ensuing harm may be caused by errors at any of these steps of the diagnostician's work, raising several questions worthy of introspection, peer review, or review for training purposes:

- Did I correctly note the patient's chief complaint and motive (reason) for consultation?
- Did I gather the necessary information from the patient's medical and other history?
- Did I correctly observe the patient? If not, what error did I commit, and what may be its consequences?
- Did I isolate all relevant manifestations and information as perceived by myself and by the patient? If not, what error did I commit, and what may be its consequences?

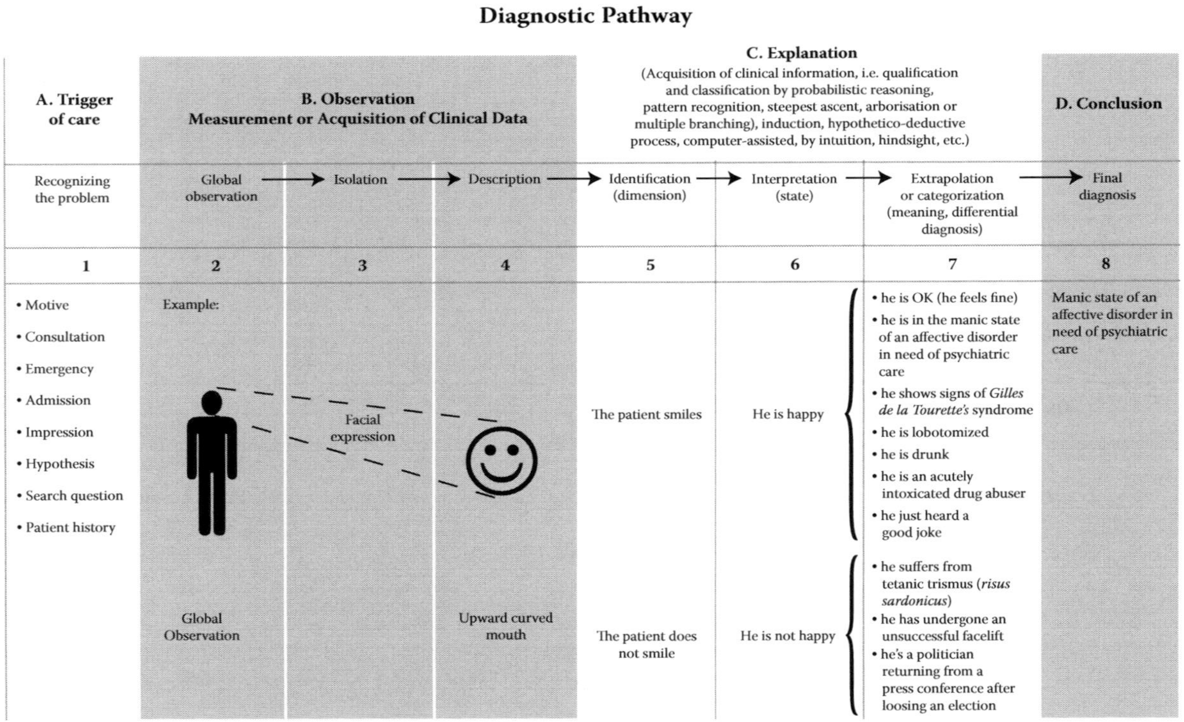

Figure 6.2 Steps in the diagnostic pathway. (Reworked from Jenicek, M., *Epidemiology*, Montreal: EPIMED International, 1995.[39])

- Did I describe them in a measurable and reproducible way? If not, what error did I commit, and what may be its consequences?
- Did I correctly perform the diagnostic test, maneuver, or procedure? If not, was this due to a sensory or motor reason, or both? Were my diagnostic maneuvers (e.g., physical examination, cardiac catheterization or rectal examination) sensory, motor, or both? At what level did lapses and mistakes occur?
- Did I use a priori knowledge of the internal and external validity of a diagnostic test, maneuver, procedure, or verbal exploration (like in psychiatry)?
- Was I or am I using tests of poor sensitivity, specificity, positive or negative predictive value, or reproducibility?
- Did I choose and correctly perform other clinical and paraclinical diagnostic tests, maneuvers, or procedures?
- If so, did I perform appropriate parallel, serial, or otherwise repeated testing where required and mandatory?
- Did I omit Bayesian reasoning in diagnosis making?
- Did I properly identify and give dimension to my observations? If not, what error did I commit, and what may be its consequences?
- Did I correctly interpret the patient's state? If not, what error did I commit, and what may be its consequences?
- Did I make a correct diagnosis, based on correct extrapolations and categorizations, giving clinical meaning to my conclusions (claim in philosophy)?
- If not, what error did I commit at this stage of the diagnostic process, and what may be its consequences?
- Did I review possible cognitive errors throughout the entire path leading to the diagnosis process?
- Is my diagnosis explicit enough in giving directions regarding the patient's treatment and further follow-up? If not, what error did I commit, and what may be its consequences?
- What might and should be corrected and avoided? Why and how?
- What should I do better next time at any step and moment of this diagnostic pathway?
- How could and should I evaluate if my corrected and better method of making a diagnosis is less error prone and if and how patients may and do benefit from it?
- Was my decision to diagnose justified?

Should some kind of "clinical reasoning root cause analysis" be considered, tried, and evaluated despite all its possible limitations?

Reviewing the Path from Diagnosis to Treatment Decisions and Orders

Although we have examined the process leading to diagnosis as a clinimetric or gnostic path with all its steps and points of potential error and ensuing harm, we may also consider another path: the one from diagnosis to treatment decision, orders, and their success or failures. For our purposes, Figure 6.3 shows treatment selection, choice, and outcomes (successes and failures) as a path that encompasses the principal or key health problem and its treatment along with the treatment of comorbidities in parallel and lateral mode paths. De Bono,[41,42] as discussed in more detail elsewhere,[6,7] expressed these ways of thinking as cutting through thinking patterns to generate new or additional concepts and ideas (lateral approach) and as multiple problem considerations (e.g., more than one treatment at a time) in pair with the vertical approach to problem solving focusing on the main health problem and its treatment.[6,7,39,40]

Treatment choice, decision, and execution and its results (and other outcomes) may be seen and analyzed as a three-step procedure starting from the treatment "baseline" or initial state followed by treatment maneuvers (e.g., medical, surgical), resulting in a "subsequent state" or various outcomes.[43,44]

Because patients may, besides the key health problem (disease), have other health problems (comorbidity) and be subject to cotreatments for comorbidities, both parallel and lateral manners of reasoning and decision making must be applied. The challenge of comorbidity was recently reviewed again by de Groot et al.[45] Comorbidity and treatment for comorbidity with ensuing outcomes are relevant at all levels of clinical exploration and care decisions.

Medical errors can occur if the clinical baseline information is wrong or incomplete, decisions to treat key morbidity and comorbidities are wrong, and expected and unexpected outcomes of are missed, poorly detected, and evaluated along with their successes and harms.

Correct or flawed fulfilling of right or wrong medical orders is the execution component of the reasoning and decision-making error.

Reviewing Decisions as Sources of Error and Harm

Generally speaking, errors in the medical decision domain may occur both in the development of decision tools and in their uses.

At the stage of their development, for example, decision analyses, trees, or tables may not be properly structured and supported by the probabilities, utilities, or evidences needed.[4] At the stage of their uses, decision tools may not be properly chosen for the health problem itself, in relation to the background information and evidence, specificity of the topic (health problem), and their

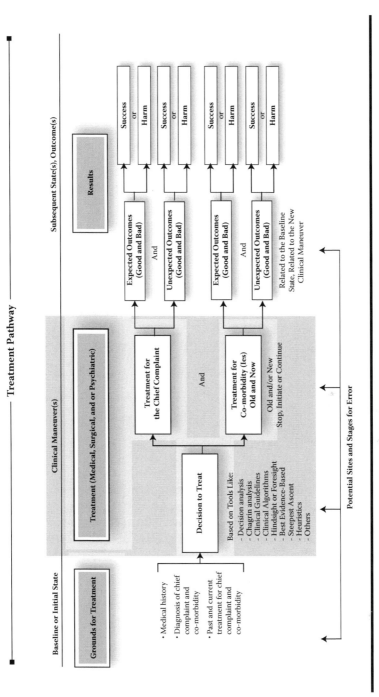

Figure 6.3 Steps in treatment pathway at which errors may occur and harm ensue.

uses in a particular setting of clinical care by a particular team using particular equipment to accomplish the task. Detmer et al.[47] remind us that heuristic situations do not facilitate decisions either. Developers and users beware.

From a management point of view, Caruth and Hadlogten[48] stress several mistakes to avoid in decision making:

- Failure to recognize the problem.
- Incorrect problem identification.
- Insufficient consideration of alternatives.
- Inadequate evaluation of risk.
- Repetitive decisions.
- Unnecessary decisions.
- Delayed decisions.
- Lack of follow-up.

How frequent such decision mistakes are in the medical domain and how much harm they produce still remains poorly known and understood. Decision tree or algorithm developers and users, beware!

Reviewing Actions as Sources of Error and Harm

We may see our errors in the acts of care as flawed actions (moments and sites for error and harm) in the execution of agreed upon surgery or invasive diagnostic or prognostic procedures including any kind of catheterization of respiratory, circulatory, digestive, urinary, and other systems or blood or tissue drawing for biopsy, cultures, and other laboratory testing.

Whereas the decision error as previously discussed is predominantly a reasoning flaw, an execution error is to a great degree of a sensory-motor nature. In other words, execution of medical orders is predominantly of a sensory-motor task as seen in surgeries, deliveries, implants of new technology devices, dental care, prosthetics, intracorporeal exploratory tools, delivery of anesthesia or emergency medicine, and intensive-care therapeutic and stabilizing maneuvers.

All activities and actions in sensory-motor skills-dependent medical specialties require the following:

- The best possible motor, tactile, visual, olfactory, and auditory skills for the task.
- Their mastery at an appropriate learning level.
- Proper disposition of caregivers (e.g., psychological, equipoise, freedom from stress, fatigue).
- Harmonized ergonomics in an advanced man–machine world.

- Unhindered communication and other interaction facilitating conditions within and between teams with complementary training, professions, cultures, philosophy, preferences, and values.

Errors may occur then due to execution flaws for any of these five reasons. Did the error occur because the physician did not have required skills? Did the physician learn the skills satisfactorily? Was the physician in a proper disposition to correctly execute required actions and maneuvers? Did the ergonomics of a given action facilitate or complicate interventions? Did the required interaction between the health professionals involved contribute to the error, or was it the main cause of error?

The *ergonomic assessment of care* and contribution to patient safety includes the following:[49]

- Involvement in the evaluation and implementation of organizational strategies that encourage patient safety incidents and near misses.
- Measurement and recording of those incidents including variables having a possible impact on patient safety.
- Giving directions to interventions aimed at improving patient safety.

For Gawron et al.,[50] the goal of human factors engineering is the optimal relationship between humans and systems. For those authors, human factors engineers design systems and human–machine interfaces that are robust enough to reduce error rates and the effect of error within the system. Root cause analysis, fault tree analysis, Petri nets, Reason's generic-error modeling system, failure mode, and effect analysis have already been used and attempted in various medical and surgical domains.

Ergonomic analyses focus on systems, that is, sets of interacting and interdependent entities in an integrated whole that includes both man and machine elements and their interrelationships. In this spirit, ergonomic analyses are also being involved in the quality movement in business and other settings. Drury[51] believes that ergonomics will prove useful in open systems and strategic issues, organization design and leadership, measurement-based operation, and appropriate uses of technologies as well as in individuals, teams, and the change process.[51] Does his "error-free manufacturing" share some common traits with "producing improvements in patient health?"

Ergonomics has been increasingly applied in the domain of health and patient safety over the past decade, but it should still be better known and used by health professionals at large and beyond the specialized health evaluation field.

Obtaining Results and Evaluating Their Impact

Errors may occur even at the evaluation level. Using or not using treatment and other health interventions, evaluated erroneously as good or bad, may harm the patient.

Several different aspects and angles of evaluation are currently used. Their meanings and purposes are complementary, however. We can expand some of Avedis Donabedian's three original concepts of evaluation (structure/process/impact or outcomes) as follows in view of various applications modified for quality assurance in healthcare,[52] health administration, care and planning, and surgery:[53-55]

- Multiangular evaluation:
  - **Soundness**: *Does it make sense?* (surgeon's knowledge and experience; evidence in its broadest sense; plausibility and its reasons)
  - **Structure**: *How is it organized?* (Are all elements of a given activity present in a clear direction and within an acceptable logical structure?)
  - **Process**: *How does it work? Does it work as desired and expected?* (Operational research or operations research is devoted in part to this type of evaluation; clinical trials and their alternatives, observational analytical studies)
  - **Impact**: *What does it do?*
  - Does it work *in ideal conditions*? This is called efficacy.
  - Does it work *in habitual conditions*? This is called effectiveness.
  - Does it work *proportionately in relation to money, time, and human and material resources spent*? This is called efficiency.
- Evolving changes in the previous evaluation.
  - **Constancy**: *Does this information change or not over time, in different places, patients, care providers, and settings?* (fluctuations of soundness, structure, process, or impact)
  - **Consistency**: *Does it make the same sense?* (Do the results follow the line of previous experience? Do they have a meaning similar to what we already know in terms of biological, contextual, and practice plausibility based on previous studies and their findings?)

Some treatment or other component of medical care may be considered sound. However, this does not automatically mean that its uses are well organized in a proper process. Its impact may be good in ideal conditions, but not necessarily in different settings, and its uses may be irrational given what we have invested in their practice. Even if such desired conditions are reached, quality of care may change over time and may acquire, during such processes, different meanings, better or worse.

Simply applying an "It works" treatment without more precision and qualification may cause harm. One of its qualities does not warrant another.

Errors in Making Prognoses

Prognosis is the prediction of being well or not over a certain period of time. Just like diagnosis, prognosis leads to considerations of further treatment and care if it is not good. To make correct decisions, more than one piece of information is usually needed:

- Both positive and negative outcomes should ideally be known in their sequence over time.
- Treatments and care moments must be detected over time.
- Alternative portraits of disease course with or without treatment and care plans should be considered wherever and whenever possible.
- Components of disease spectrum and gradient leading to or being sites responsible for a good or bad prognosis should be known as well as possible.
- Effectiveness of interventions altering prognosis should be known in advance. It may not be the same as the one altering the risk of developing the disease.
- Distinctions between risk factors and prognostic factors (sometimes called hazard factors) are important.
- Distinctions between risk characteristics and hazard (prognostic) characteristics are sometimes blurred.

From the risk domain, we make distinctions between risk markers (which cannot be modified like age and sex of the patient) and risk factors (characteristics which can be modified to minimize the occurrence of disease). The control of risk factors lowers the probability of developing disease. Similarly, knowledge of prognostic markers and prognostic factors will lead us to better understand what is and is not effective to improve patient prognosis (i.e., the disease course) once it is established.[39,40]

An additional point is that some prognostic factors may be more or less important than similar risk factors. For example, smoking is a more important risk factor than a prognostic one. Alcohol abuse may have an equal role in the development of liver disease and in its further course (prognosis) once the diagnosis of alcoholic hepatitis or cirrhosis is established.

In this sense, for example, five-year survival rates as measures of prognosis may not always prove satisfactory for all correct care decisions. They remain, however, one of the valuable leads for better management of patient prognosis. Using prognostic trees to detect sites of possible error is limited given that, in contrast to decision trees, currently available prognostic trees do not include interventions and their outcomes and utilities. By looking at prognosis this way, errors in prognosis and subsequent decisions in caring for the patient who has already developed a disease may be due or explained by mistakes (not making necessary distinctions and related overencompassing care decisions) connected to any point previously mentioned.

Errors in prognosis may occur if prognosis for the patient, prognosis for the key problem, and prognosis taking into account variations in comorbidity(ies) are confounded. If we conclude that "prognosis is good," is it for and of what of the above?

Evidence-based medicine reminds us that the best evidence should be used in conjunction with patient preferences and values and with the setting of care among others. Using even the best evidence on prognosis also requires some kind of *setting fit*:

- *Fit of patient characteristics* with those who served as the basis for prognostic information.
- *Research fit*: The purpose of reference prognostic studies should be in line with the patient problem such as methods of follow-up, medication, and surgery used.
- Medical *care fit:* The setting, rules, and everyday run of care should be close to the ones of the patient.
- *Caregiver fit:* Physicians, nurses, and other healthcare professionals should be comparable to their competencies, ways of practice, and general "hospital culture."
- *Prognostic fit,* as already discussed in the previous paragraphs of this section.

Errors in prognoses and possible harm from them may also result from discrepancies between reference values and information and their application in a specific clinical environment.

Prognosis, like any step in the clinical care that precedes it, may be erroneous and generate other errors if it is poorly defined, poorly described, and poorly analyzed and if it generates poor conclusions, recommendations, and further medical care orders.

Follow-up, Surveillance, Forecasting-Related Errors

A follow-up of error events and harm, needless to say, should go beyond the original statement and interpretation of medical error event.

In the medical error domain, surveillance is limited so far to serious cases in high-risk environments. Should medical error surveillance and forecasting follow the general rules of epidemiological surveillance,[56] ideally as a systematic and continuous collection, analysis, and interpretation of data, closely integrated with the timely and coherent dissemination of results and assessment to those who have the right to know so that action can be taken? Surveillance should be performed irrespective of the gravity of the problem at a given moment; therefore, it should fluctuate in time.

Without surveillance of medical error (its frequency and high occurrence of its causes permitting and justifying the endeavor), better causal research and evaluation of the consistency and constancy of error events and their prevention are limited.

Epidemiological forecasting is a method of estimating what may happen in the future based on the extrapolation of existing trends.[56] Its subjects of interest may be clinical, demographic, community based, or epidemiological, among others. Even in the forecasting domain, initiatives should take into account sufficient numbers of events that make forecasting possible beyond the hypothetical level.

To make surveillance and forecasting more meaningful in the medical error domain, simple registries of cases without possible causal factors relevant for prevention and control may prove of limited value. *Deductive build-up* of medical error related variables, data, and information may prove more feasible and rational than *inductive exploration*, an "all azimuths" multiple data collection stemming from the fear that "nothing must be forgotten!" Medical error reporting in the United States, however, remains complex.[57]

Only more recently, probabilistic risk analysis (discussed also in Chapter 5) was proposed as a tool to make additional predictions based on few or rare cases in the framework of medical error, harm, and sentinel events.[58,59]

Follow-up, surveillance, and forecasting in the medical error domain merit further development and wider application and uses. In an epidemiological sense, isn't medical error another "disease" that should be properly detected, treated effectively, and prevented? This would be much harder without proper surveillance and forecasting, so useful in other domains of clinical and field (community) epidemiology.

Conclusions

The model needed for an evaluation of sources of error is very close to the components and sources of teaching and learning skills in surgery:[54,55] errors and harm may originate from faulty fundamental assets that need to be acquired (knowledge, attitudes, skills) and from misuses of passive and active teaching tools as well as from methods of communication.

Current discussion and experience focus mainly on a single or a particular type of error. In addition to this fundamental approach, should we consider in the future a kind of *meta-evaluation of human error* across its spectrum and gradient including various medical errors of interest? In both approaches, medical error by nature falls into three interlinked categories:

■ It may be the product of the faulty knowledge, attitudes, and skills of the health professional. Unsound evidence is used, or the best sound evidence isn't. Evidences used in medical understanding and decisions may be unsound, poorly structured, and executed in an inappropriate process, with an unclear impact in terms of effectiveness, efficacy, and efficiency, not constant and inconsistent in time and space.
■ All the following may occur at any step of medical care as described earlier in this chapter:
  − The entry of the patient under care may be questioned.
  − Patient interviews, both medical and psychiatric, may be flawed.
  − Physical and paraclinical examinations may yield wrong data and information.
  − Risks of expected health problems in the patient may be misestimated.
  − Diagnosis and differential diagnosis may be wrong, and diagnosis of other problems (comorbidity) may be wrong or incomplete.
  − The plan of care may be too narrow or too broad or well or poorly defined.
  − Treatment modalities may be misdirected or incomplete, or comorbidity may be omitted; prognosis may be wrong, omitting again comorbid states and the occurrence of additional new health problems, their treatment, and interactions with the main problem.
  − Follow-up may not be long enough or complete enough to control long-term effects and needs.
■ Sources of error at the individual level may include inadequate knowledge, attitudes, and skills as well as poor critical thinking and argumentation. Technology problems may include both their design and uses in a good or bad physical environment of technology at work. In addition, systems may be at fault: the operator may interact inappropriately with the machine, members of the healthcare providing team may poorly interact among themselves, and communication at any level of any system may be compromised.

Figure 6.4 illustrates this concept in a simplified matrix form.

This kind of mapping of the spectrum of medical error and the gradient of the harm it produces should prove an additional asset in planning prevention and control of the medical error problem across clinical and community care.

Moreover, in light of this chapter, what should we do to analyze the individual human component in a structured way? Claims of any kind at any level of the analysis and evaluation of the process and results of care should be based

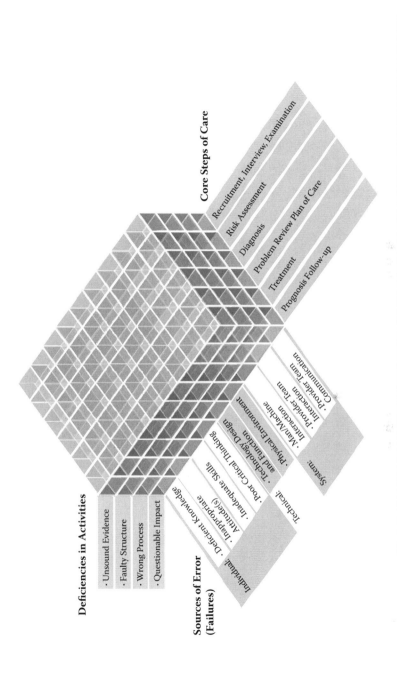

Figure 6.4 Distribution of medical errors according to the level of acquired professional competency, steps in clinical care, and desirable quality attributes of medical interventions.

on a structured argumentation in which conclusions are based on clinically and logically meaningful grounds, backing, warrants, qualifiers, and rebuttals supporting the overall understanding of the problem under study.

Therefore, loci of potential error and harm and needed attention across the stages of clinical care are as follows:

1. Define the kind of error possibly associated with the harm observed.
2. Consider the human (operator) error first before expanding the inquiry at the system level.
3. Examine the step-by-step clinical care process for possible locations of mistakes leading to the error event.
4. Consider both flaws in reasoning and decision making and sensory-motor actions.
5. See if the error event is due to the patient's entry under care: the chief complaint and reason for consultation or hospitalization was correct, missed, or misdirected. Incorrect care may follow.
6. Identify, if possible, the precise stage in the clinimetric process that is an incorrect step responsible for further incorrect decisions in patient care: observation, qualification, categorization, others?
7. Were the rules of hypothetico-deductive thinking, proper pattern recognition, algorithms, or other diagnostic guidelines well chosen given the case and circumstances?
8. Verify possible faults in making differential diagnosis and final diagnosis and their completeness in view of multiple morbidities (comorbidity).
9. Clarify how treatment decisions and orders were justified and made: check errors in treatment guidelines or algorithm choices, decisions based on heuristics, decision analysis, others?
10. Analyze and explain actions in care as potential loci for mistakes (flaws in reasoning and decision making) and slips and lapses (mostly of a sensory-motor nature).
11. Try to understand how prognosis was made: did it take into account relevant prognostic markers and factors, comorbidity and cotreatments for comorbidity, patient characteristics, and clinical and community care settings? What are the decisions in light of the wrong prognosis that may lead to future potential errors in care and harm?
12. Is the proposed patient follow-up and surveillance justifiable by all preceding steps, and how does it fit beyond single-case forecasting for this type of health problem and the healthcare related to it?
13. Analyze the system if the operator's errors do not explain satisfactorily the error event of interest. If not, consider system analysis.

Should we see this approach to clinical error understanding as a kind of root cause analysis at the human operator level?

Each and every human participant in the "system" that produces error and harm may contribute to them at any stage and point in medical care through the following:

- Flaws in initial hypothesizing, questions to answer, and objectives for further action.
- Flaws in perception of bodily, mental, and behavioral manifestations.
- Flaws in interpretation of what was seen and its coupling with good or bad supporting background evidence.
- Flaws in making conclusions like diagnosis or treatment orders.
- Flaws in execution or actions, mostly of a sensory-motor nature.
- Flaws in the evaluation of actions.
- Flaws in patient or community follow-up.
- Lack of feedback evaluation of flaws and their correction.

Hence, human medical error may be caused by flaws in argumentation and critical thinking when producing, interpreting, using, and evaluating evidence and the ways it was used.

Ideally, before any analysis of system error is attempted, human error as described in this chapter should precede or be part of any mutifactorial analysis within the system involving all individuals who are part of such systems and who interact with all additional operational, technological, technical, environmental, and other physical, biological, and social components of systems that ultimately produce error and ensuing harm.

However imperfect our understanding of the human contribution to error by perception, reasoning, decision making, and action may be, can we leave it as a black box within the system concept and its analysis in lathology? We believe not.

References

1. Fabri PJ, Zayas-Castro JL. Human error, not communication and systems, underlies surgical complications. *Surgery,* 2008;**144**:557–65.
2. Wilson R McL, Harrison BT, Gibbert RW, Hamilton JD. An analysis of the causes of adverse events from the Quality in Australian Health Care Study. *MJA,* 1999;**170**:411–5.
3. Makeham MAB, Dovey SM, County M, Kidd MR. An international taxonomy for errors in general practice: a pilot study. *MJA,* 2002;**177**:68–72.

4. Delbanco T, Bell SK. Guilty, afraid and alone—struggling with medical error. *N Engl J Med,* 2007;357(17):1682–3.
5. Jenicek M, Hitchcock DL. *Evidence-Based Practice. Logic and Critical Thinking in Medicine.* Chicago: American Medical Association Press, 2005.
6. Jenicek M. *A Physician's Self-Paced Guide to Critical Thinking.* Chicago: American Medical Association Press, 2006.
7. Jenicek M. *Fallacy-Free Reasoning in Medicine. Improving Communication and Decision Making in Research and Practice.* Chicago: American Medical Association Press, 2009.
8. Toulmin SA. *The Uses of Argument.* Updated Edition. Cambridge, UK: Cambridge University Press, 2003. (First Edition: Idem, 1958.)
9. Toulmin S, Rieke R, Janik A. *An Introduction to Reasoning.* 2nd Edition. New York: Collier Macmillan Publishers, 1984.
10. Jenicek M. Towards evidence-based critical thinking medicine? Uses of best evidence in flawless argumentations. *Med Sci Monit,* 2006;12(8):RA149–RA153.
11. Reason JA. *Human Error.* Cambridge, UK: Cambridge University Press, 1990.
12. Dictionary.com. *Mistake—10 dictionary results.* Available at: http://dictionary.reference.com/browse/mistake, retrieved February 10, 2009.
13. Merriam-Webster OnLine. *Mistake.* Available at: http://www.merriam-webster.com/dictionary/mistake[2], retrieved February 10, 2009.
14. Merriam-Webster OnLine. *Error.* Available at: http://www.merriam-webster.com/dictionary/error, retrieved February 10, 2009.
15. Dowden B. *The Internet Encyclopedia of Philosophy. Fallacies.* Available at: http://www.iep.utm.edu/f/fallacy.htm, retrieved October 31, 2006.
16. Henriksen K, Kaplan H. Hindsight bias, outcome knowledge and adaptive learning. *Qual Saf Health Care,* 2003;12(Suppl II):ii46–ii50.
17. Chapter 21. Scientific decisions in choosing groups. (pp. 458–499) and 22. Scientific decisions for data and hypotheses (pp. 500–532) and elsewhere in: Feinstein AR. *Clinical Epidemiology. The Architecture of Clinical Research.* Philadelphia: W.B. Saunders Company, 1985.
18. Absence of bias. Section 8.8.6 (pp. 206–209) in: Jenicek M. *Foundations of Evidence-Based Medicine.* Boca Raton, FL: Parthenon Publishing Group, a CRC Press Company, 2003.
19. Lilford RJ, Mohammed MA, Branholts D, Hofer TP. The measurement of active errors: methodological issues. *Qual Saf Health Care,* 2003;12 (Suppl II):ii8–ii12.
20. Absence of bias. Section 6.8.6 (pp. 180–183) in: Jenicek M. *Epidemiology. The Logic of Modern Medicine.* Montreal: EPIMED International, 1995.
21. Redelmeier DA. The cognitive psychology of missed diagnoses. *Ann Intern Med,* 2005;142(2):115–20.
22. Paradis C. Bias in surgical research. *Ann Surg,* 2008;248(2):180–8.
23. Croskerry P. Cognitive and affective dispositions to respond. Chapter 32, pp. 219–227 (Table 32-2) in: Croskerry P, Cosby KS, Schenkel SM, Wears RL. *Patient Safety in Emergency Medicine.* Philadelphia: Wolters Kluwer | Lippincott Williams & Wilkins (Health), 2009.
24. The Nizkor Project. *Fallacies.* Available at: http://www.nizkor.org/features/fallacies/, retrieved October 10, 2006.

25. Thompson B. *Bruce Thompson's Fallacy Page.* Available at: http://www.cuyamaca. net/bruce.thompson/Fallacies/intro_fallacies.asp, retrieved November 1, 2006.

26. Curtis GN. *Fallacy Files.* Available at: http://www.fallacyfiles.org/taxonomy.html, retrieved October 30, 2006.

27. *EvoWiki Encyclopedia of Fallacies.* Available at: http://wiki.cotch.net/index.php/ List_of_fallacy_pages, retrieved March 12, 2007, and October 29, 2006.

28. Downes S. *Stephen's Guide to the Logical Fallacies.* Brandon, Manitoba, Canada, 1995–1998. Available at: http://www.intrepidsoftware.com/fallacy/welcome.php and related sites, retrieved December 6, 2004. See also http://www.assiniboinec. mb.ca/user/downes/fallacy.

29. Damer TE. *Attacking Faulty Reasoning. A Practical Guide to Fallacy-Free Arguments.* Fifth Edition. Belmont, CA: Thomson Wadsworth, 2005. (Second Edition: Belmont, CA: Wadsworth Publishing Company, A Division of Wadsworth, Inc., 1987.)

30. Holt T. *Logical Fallacies. An Encyclopedia of Errors of Reasoning.* Available at: http:// www.logicalfallacies.info/, retrieved October 10, 2006.

31. Anon. *Logical Fallacies and the Art of Debate.* Available at: http://www.csun. edu/~dgw61315/fallacies.html, retrieved February 17, 2007.

32. Lindsay D. *A List of Fallacious Arguments.* Available at: http://www.don-lindsay-archive.org/skeptic/arguments.html, retrieved February 15, 2007.

33. Lindsay D. *A List of Fallacious Arguments.* Available at: http://www.don-lindsay-archive.org/sceptic/arguments.html, retrieved March 25, 2009.

34. Changing Minds.org. *Fallacies: alphabetic list (full list).* Available at: http://changing-minds.org/disciplines/argument/falalcies/falalcies_alpha_htm, retrieved March 25, 2009.

35. BambooWeb Dictionary. *List of Cognitive Biases.* Available at: http://www. bambooweb.com/articles/L/i/List_of_cognitive_biases.html, retrieved March 4, 2009.

36. Gunderman RB. Biases in radiologic reasoning. *AJR,* 2009;192(March):561–4.

37. Feinstein AR. Clinical biostatistics LVII. A glossary of neologisms in quantitative clinical science. *Clin Pharm Ther,* 1981(Oct);**30**(4):564–77.

38. Feinstein AR. *Clinimetrics.* New Haven, CT: Yale University Press, 1987.

39. Jenicek M. *Epidemiology. The Logic of Modern Medicine.* Montreal: EPIMED International, 1995.

40. Jenicek M. *Foundations of Evidence-Based Medicine.* Boca Raton, FL: Parthenon Publishing Group, a CRC Press Company, 2003.

41. de Bono E. *Edward de Bono's Thinking Course.* Harlow, Essex: BBC Active, 2006. (Lateral and parallel thinking)

42. de Bono E. *Lateral Thinking. A Textbook of Creativity.* London: Penguin Books, 1990 (reprint from Ward Lock Education, 1970).

43. Feinstein AR. *Clinical Biostatistics.* St. Louis: C.V. Mosby, 1970.

44. Feinstein AR. *Clinical Epidemiology. The Architecture of Clinical Research.* Philadelphia: W.B. Saunders Company, 1985.

45. de Groot V, Beckerman H, Lankhorst GJ, Bouter LM. How to measure comorbidity: a critical review of available methods. *J Clin Epidemiol,* 2003;**56**:221–9.

46. Decision analysis and decision making in medicine. Beyond intuition, guts and flair. Chapter 13, pp. 341–378 in Ref. 40.

47. Detmer DE, Fruback DG, Gassner K. Heuristics and biases in medical decision-making. *J Med Educ,* 1978;**53**(Aug):682–3.

48. Caruth DL, Hadlogten GD. Mistakes to Avoid in Decision Making. *Innovative Leader,* 2000;9(7). Available at: http://www.winstobrill.com/bril001/html/article_index/articles/451-500/article477_body..., retrieved February 2, 2009.

49. Edworthy J, Hignett S, Hellier E, Stubbs D. Patient safety. *Ergonomics,* 2006;49(5–6):439–43. See also other related articles in this issue.

50. Gawron VJ, Drury CG, Fairbanks RJ, Berger RC. Medical error and human factors engineering. *Am J Med Qual,* 2006;**21**(1):57–67.

51. Drury CG. Ergonomics and the quality movement. *Ergonomics,* 1997;40(3):249–64.

52. Basic approaches to assessment: Structure, process, and outcome. Chapter Three, pp. 77–128 in: Donabedian A. *Explorations in Quality Assessment and Monitoring. Volume I. The Definition of Quality and Approaches to Its Assessment.* Ann Arbor, MI.: Health Administration Press, 1980.

53. The Justification of Surgery. Chapter 4 (pp. 95–162) in: Donabedian A. *Explorations in Quality Assessment and Monitoring. Volume III. The Methods and Finding of Quality Assessment and Monitoring: An Illustrated Analysis.* Ann Arbor, MI: Health Administration Press, 1985.

54. Jenicek M. Clinical case reports and case series research in evaluating surgery. Part I. The context: General aspects of evaluation applied to surgery. *Med Sci Monit,* 2008;**14**(9):RA133–143.

55. Jenicek M. Clinical case reports and case series research in evaluating surgery. Part II. The content and form: Uses of single case reports and case series research in surgical specialties. *Med Sci Monit,* 2008;**14**(10):RA149–RA162.

56. *A Dictionary of Epidemiology.* Fifth Edition. Edited by M Porta. Oxford: Oxford University Press, 2008.

57. Karlsen AK, Hendrix TJ, O'Malley M. Medical error reporting in America: A changing landscape. *Q Manage Health Care,* 2009;**18**(1):59–70.

58. Alemi F. Probabilistic risk analysis is practical. *Q Manage Health Care,* 2007;16(4):300–10.

59. Alemi F. Tutorial on discrete hazard functions. *Q Manage Health Care,* 2007;16(4):311–20.

Chapter 7

Prevention, Intervention, and Control of Medical Error and Harm: Clinical Epidemiological Considerations of Actions and Their Evaluation

Executive Summary

If we accept the dual theory of error and harm generated either by human (individual) or system error, our measures to control them must address individuals, systems or both. At any level of their prevention (i.e. primordial, primary, secondary, tertiary or quaternary) controlling or eliminating them may be viewed as putting into practice experimental methodology to understand and resolve the cause-effect relationship between the origins of error and harm and between error and harm as a cause of its consequences (outcomes).

Measures undertaken in lathology must be founded on clear objectives because their evaluations in terms of their relevance, effectiveness or impact are meaningful only as much as the clarity of intervention objectives.

Any activity, operation, or program in clinical or community care should show changes in occurrence and/or severity and demonstrate measurable and meaningful human biological or social (environmental) impact. The evolution of such changes should be followed over time.

However, medical error and harm protection refers to measures used to avoid or minimize exposure to potential causes of error and the harm they generate. Defined as such, it is a public health measure subject to health protection rules and experience in other domains.

Broader and better-defined experience in freedom from medical error and harm as a health promotion activity still remains limited. If health promotion activities should provide health professionals and their patients, institutions and communities increased control over their error-free clinical and community care and the resulting improvements and maintenance of health, in lathology it should mean strengthening the ability of health professionals and their patients to avoid and remain free from error and its consequences. This may be challenging if promotion activities are (as they should) to be translated in operational terms of activities of interest.

As in any other health domain, health programs in lathology have a structure, process and impact to follow. At the individual (human) level, knowledge, attitudes and skills all contribute to good or bad outcomes and other results of activities expected to reduce errors and harm, and their consequences.

Given the nature of problems in lathology like frequency, ethics, access to data or completeness and complexity of information, interventions and their evaluation must go beyond experimental methodology: Quasi-experimental, non-experimental (observational) methods, natural experiments, before-after studies, case studies must not be forgotten despite their limitations compared to the experimental approach.

Healthcare failure mode and effect analysis (HFMEA) and systematic reviews of evidence, where the nature of the problem and information available allow, are additional tools to be considered in evaluating the impact and success of activities devoted to minimize medical error and its consequences.

Thoughts to Think About

Physician, heal thyself.

—The Bible, Luke 4:23

Shouldn't we correct our own errors first before pointing to the errors committed by others? I see no dignity in persevering in error.

—Robert Peel, 1833

Analyze. Think, think, think…. We are so accustomed to faulty states of mind that it is difficult to change with just a little practice. Just a drop of something sweet cannot change a taste which is powerfully bitter. We must persist in the face of failure.

—His Holiness the Dalai Lama, 2002

Any error does not become a mistake until you refuse to correct it.

—E. C. McKenzie, 1980

Never mistake motion for action.

—Ernest Hemingway, 1898–1961

"Do-so" is more important that "say-so."

—Peter Seger, 1919–

Thought is the blossom; language the bud; action the fruit behind it…. All life is an experiment. The more experiments you make the better.

—Ralph Waldo Emerson, 1803–1882

Mistakes are a fact of life. It is the response to error that counts.

—Niki Giovanni, 1970

The real problem isn't how to stop bad doctors from harming, even killing, the patients. It's how to prevent good doctors from doing so.

—Atul Gawande, 1965–

The actions we take to understand and correct medical error and harm and to evaluate our success in these matters are perhaps even more important than some (not all) reflections and understandings that lead us to them.

—Author unknown

If the search for causes of medical error is as challenging as we have seen in Chapters 5 and 6, preventing and controlling medical error is even more so. Without an understanding of the causes and webs of causes underlying any health problem, it is difficult, if not impossible, to prevent and control the health problem in question. Medical error and its consequences (harm) do not escape such a rule.1 This chapter is about corrective actions applied to human error and harm prevention and control in a manner acceptable to epidemiologically minded health professionals. It is a challenge, and compromises must be expected.

Introductory Comments, Interventions in the Medical Error Domain

Medical error events may have complex webs of causes considered in lathology to be spatial chains of events and happenings leading to error and harm. A physician may prescribe the wrong drug, a pharmacist may misread a prescription, a clinical nurse may not note any discrepancy in prevalent rules; a *system failure* would then be behind any such unfortunate event. Besides incorrect technologies related to surgical, diagnostic, and exploratory instruments and therapeutic technical tools (e.g., pacemakers, medicated stents in cardiology or artificial joints and bone transplants in orthopedic surgery that may break or dysfunction), at any point in the "system" the physicians, nurses, pharmacists, technicians, or other human reasoners and decision makers can always fail. Preventing and controlling medical error means paying equal attention to the entire system and to all humans involved in it. Within the system failure, there is always a *human error caused by an individual or group of individuals,* especially those on the "sharp end" (i.e., decision makers and executors) of a clinical task.

If system error and its control are in the spotlight, the ensuing evaluation will generally not be simple. Both modern health problems and practice are complex. They require customized uses of methodology in the general medical domain[2] and in lathology itself.[3-5]

With interventions in mind, the epistemology of patient safety research now has a more solid methodological basis[6] that we can place in the evaluation context of actions in the medical error domain as explained in this chapter. Past and current experience[7] and available methodologies may be beneficial in most medical specialties such as initiatives in various specialties like psychiatry,[8] obstetrics, and gynecology.[9-11] Given the seriousness of the medical error problem and imperatives to act rapidly and efficiently, we cannot always wait for full knowledge of medical error causes to which we must add other compromising considerations.

Current recommendations, guidelines, algorithms, and procedural rules are based on the following:

- The best intentions, feelings, motivation, and hunches.
- Extrapolation of successes and failures from past experience stemming largely from before–after observations.
- Biological, physical, and technological plausibility.
- Orders from expert authorities in positions of administrative, regulatory, and executive authority.
- Ideally, on epidemiologically sound interventions that are effective, efficient, and efficacious in as precise and measurable terms as possible.

And because we feel that things cannot be left as they are and that something must be done right away, this chapter examines these five points. They apply to protection from medical error, prevention of medical error, and even promotion of medical care practices guaranteeing freedom from medical error and its consequences.

Basic Definitions, Concepts, and Strategies of Intervention in Lathology

Ideally, cause–effect relationships and anything we do to control medical error are subject to experimental methodology, however limited our actions may be due to medical ethics or a fortunately low frequency of medical error cases. Cause–effect relationships and their management apply not only to causes of error but also to causes of the medical error problem improvement or failure.

Contrary to observational analytical studies, *studies in experimental epidemiology and other types of medical care assessment* rely on an investigator's intentional alteration of one or more independent factors and controlling of other study conditions to analyze the effects of doing so. Phase III or phase IV clinical trials, community field trials, or randomized or otherwise controlled evaluations are experimental studies.[12]

Figure 7.1 illustrates the experimental approach to evaluating cause–effect relationships. Clinical trial in its classic form is experimental research focused on the causes of improvement in disease occurrence (e.g., the protective effect of a vaccine) or disease course improvement once it occurs. An experimenter raises such questions and hypotheses, selects all independent (causes) and dependent (disease frequency and nature) variables, chooses individuals eligible for such a trial, decides who will be subject to the intervention of interest (drug, surgery, other type of care) and who will not, and decides on the type of such an assignment. Randomization, however desirable it may be from a scientific standpoint, is not always possible; compromises must often be made.

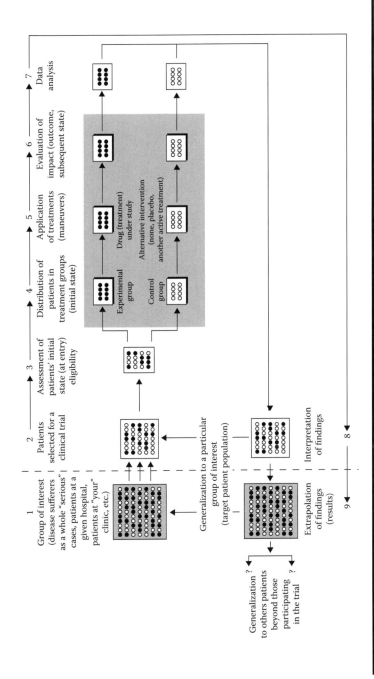

Figure 7.1 Organization and path of an experimental study. Trials, health programs, other interventions. (Modified and redrawn from Jenicek, M., *Foundations of Evidence-Based Medicine*, Parthenon Publishing Group Inc. [CRC Press], Boca Raton, FL, 2003.)

Study success depends on how a selected group of patients is representative of the target population and of the health problem itself, on how both (or all) groups compared are equally followed and assessed with regard to their initial state, on the application of therapeutic or preventive maneuvers, on ensuing outcomes and other results of interventions, and on how such trial results may be extrapolated to the target population and generalized to a desirable and realistic degree.

Two Complementary Strategies: Human Error and System Failures

As already discussed regarding causal considerations of error and harm, interventions to prevent and control error and harm may also be centered on corrections of human error and on the correction of systems that may produce it.[13] These strategies are neither contradictory nor exclusive.

In the area of *individual human error*, we pay attention to the correction of flaws in critical thinking and decision making of individuals (mostly operators), their knowledge, attitudes, intellectual and sensory-motor skills, physical and mental disposition, and fitness to be part of a system (ergonomic considerations). In regard to *system-produced error and harm*, we consider them to be the result of accumulated failures of various human, chemical, physical, technological, and environmental factors in their nature, constellation, and interaction. Following numerous antecedent and latent factors is the "sharp end" of the healthcare system consisting of institutional context, organization and management, work environment, care team, individual team member, task, and patient.[14,15] Tactics to reduce errors and mitigate their adverse effect consequently include reducing the complexity of tasks and systems, optimizing information processing, using automation and constraints, and mitigating unwanted effects of change.[3,8] Various individuals who are a part of the system may commit their own human errors, which contribute to the cumulative effect of the system in the form of webs of causes and consequences to be understood, prevented, and controlled once they occur. System error considerations are prominent as strategies to prevent and control error for both medical[16,17] and legal[18] understanding.

The balance between human and system error may vary from one specialty to another. In medicine, clinical pharmacology, and nursing care, system errors must be considered. In surgical specialties, human error may be prominent among other components of the surgical care system. However, any error may be thought of as a system error with varying proportions of human and other system components depending on the workplace, medical specialty, and setting of clinical and community care.[19-21]

Evaluation of Activities in Lathology

Evaluation (in medical lathology as well) is a process that attempts to deter-mine as systematically and objectively as possible the relevance, effectiveness, and impact of activities in light of their objectives.[12] Relevance, effectiveness, activities and their impact, and objectives of activities and their evaluation must be precisely formulated. Any and all ensuing activities and their success rely on it.

Evaluation is the systematic assessment of the operation or the outcomes of an activity, program, or policy compared with a set of implicit and explicit standards as a means of contributing to the improvement of the activity.[22] In the medical error domain, it may be seen as a process that attempts to determine as systematically and objectively as possible the relevance, effectiveness, and other characteristics of activities, programs, and policies to prevent and control medi-cal error and its consequences as well as the impact of such activities, programs, and policies in light of their objectives. Evaluation is crucial in all domains of intervention to control the medical error problem.

Evaluation methods include input, output/performance, impact/outcomes, service quality, process, benchmarking, organizational effectiveness, and other qualitative and quantitative strategies.[23]

In *medical error evaluation programs* as well, the main questions to answer are as follows:[24]

- What are the nature and basic characteristics of the program to be evaluated?
- What justifies a new program?
- Are there feasible and effective interventions?
- Do we know the target population of the program?
- How well was the intervention implemented? Was the intervention effec-tive enough in view of the expected goals, objectives, and benefits?
- Is this surgical activity sound and effective, and does it remain that way over persons/time/place (settings), runs, and fluctuations?
- Is the cost–effect ratio of the program reasonable?

In a constant flux between social sciences and health sciences, the meth-odology of evaluation in social sciences is adopting some trends from health sciences like evidence-based problem solving and decision making, meta-analysis and systematic reviews, and case studies in a wider context. In the opposite direction, health sciences are adopting methodological developments and experience from psychology, operational research, manufacturing, and transportation.

Control of Medical Error and Harm

In its broadest sense, *control (of medical error and its sequels)* means any activity, operation, or program in clinical and community care to reduce the occurrence, severity, and human and environmental (human, biological, social) impact of medical error and to follow its evolution over time (surveillance and further action if necessary).

Prevention of Medical Error and Harm

Prevention in lathology concerns both medical error and the harm it produces. Error and harm may have distinct causes and consequently distinct manners of intervention. The following multiple levels of prevention[12] are worthy of attention.

Primordial prevention,[12] a more recent entity, means the control of conditions, actions, and measures that minimize hazards to health and hence inhibit the emergence and establishment of processes and factors (environmental, economic, social, behavioral, cultural) known to increase the risk of disease. Primordial prevention is accomplished through many public and private health policies and intersecting action.[12] Focusing on the exposure to potential causes, it may be seen as a part of primary prevention (no disease involved yet).

Primary prevention of medical error and medical error produced harm means controlling the incidence (occurrence of new cases over a certain period of time) of medical error or medical error-generated harm.[25] It is generally a task of primary care, family medicine, and public health.

Secondary prevention of medical error and medical error-produced harm means controlling the prevalence (occurrence of all cases at one moment or over a certain period of time) of an error event. Given that the prevalence of error or harm is a function of its average duration and incidence,[25] secondary prevention applies to the control of medical error-induced harm rather than to the medical error itself. The control of disease duration belongs more to clinical medicine; in preventive medicine, sector "screening for disease (or for medical error and harm)" remains an important tool.

Tertiary prevention of medical error and error-produced harm means controlling the gradient and severity of disease and its long-term impact with regard to disability, impairment, and handicap. Minimizing suffering and maximizing potential years of useful life[12] is also its objective. Tertiary prevention focuses exclusively on the error-produced harm, not its causes.

Quaternary prevention consists of actions that identify patients at risk of over-diagnosis or overmedication and that protect them from excessive medical intervention—actions that prevent iatrogenesis.[12] Quaternary prevention of error and harm may prove to be of interest for operational research. This level, meaning, and sense of prevention may apply to both medical error and error-generated harm applications if supported by further evidence.

Specific epidemiological methods and techniques used in disease prevention in general are described in the epidemiological, public health, and evidence-based medicine literature.[25-28] Prevention and control in lathology are still often confused and grouped together. They should not be. Their objectives and methodologies are different.

Protection of Freedom from Medical Error and Harm

Medical error and its generated harm protection refers to measures used to avoid or minimize exposure to error and to the harm it generates. Similarly, disease protection at large means activities and programs to prevent exposure to noxious hazards, be it physical (e.g., occupational and other factors), chemical (e.g., air pollution), or biological (e.g., infections). In lathology, human and physical flaws in system-generating error are hazards from which patients should be protected. In addition to the hazards that are already known, activity also includes potential and emerging error threats to patient health.

Promotion of Freedom from Medical Error and Harm

Health promotion related to error (both hospital or extrahospital medical care and community health practices producing or generating error) based on the general definition of health promotion[12] is a process enabling health professionals and their patients and communities to increase control over their error-free clinical and community care and the resulting improvements and maintenance of health. Transposed into lathology, freedom from medical error promotion should mean strengthening the ability of health professionals and their patients to avoid and remain free from error and its consequences.

Broader experience in freedom from medical error promotion is sorely needed.

Basic Angles of Evaluation in Lathology: Structure, Process, Outcomes, and Other Subjects to Evaluate

What should be evaluated in the domain of medical error and harm regarding interventions like prevention or control of medical error and harm? We partly commented on Avedis Donabedian's concepts[29-30] (modified here by us) together with examples of application[31-33] in Chapter 6, Section 6.55. Some of these concepts apply to the evaluation of any activity, but they are especially important in the domain of evaluating health programs and other interventions.

In summary, subjects of evaluation may be roughly divided into "snapshot" evaluations and "evolving changes" of the former:

1. In ***snapshot evaluations***, we are interested in **soundness** (*Does it make sense?*), **structure** (*How is it organized?*), **process** (*How does it work and does it work according to our expectations?*) and **impact and outcomes** (*What does it do, what were the results?*).

2. These snapshot evaluations may evolve in time if health programs and interventions are repeated from one time, place, or person to another. ***Evolving changes*** should complete the picture based on the **constancy** (*Does it always produce the same outcome?*) of the health program and intervention results, i.e. of their structure, process, and/or impact, and their **consistency** (*Does it make the same sense?*) with previous results and experience in similar and other programs.

This model, already more detailed in Chapter 6, Section 6.5.5 was originally developed for health administration and evaluation of health services and quality of care. With other considerations in Sections 7.4 and 7.5 that follow, it applies equally well to evaluation in medical error and harm interventions and actions evaluation. At the moment, however, this type of evaluation is either in its infancy or it is not always explicit from one study to another. The reader must often figure it out for himself or herself.

What Should Be Evaluated at the Individual Level: Knowledge, Attitudes, and Skills

At the individual level, errors occur if the operator's knowledge is deficient and insufficient, if his or her attitudes are not appropriate for the task, or if his or her skills are inadequate for the successful execution of the task. Their improvement is often an explicit or tacit tool to prevent and control medical error.

In the context of Boolean taxonomy,[34-36] three qualities of each individual, potential producer of error and harm as well, as its preventer and controller are then as follows:

- What do I know about the means of interventions, programs, and activities to prevent and control a health problem? Is my knowledge actively or passively acquired?

■ What are my attitudes (values and judgment of them) towards preventive and control measures? Do I consider them worthy of consideration and why? How do I foresee their success or failure?

■ Am I equipped, trained and experienced enough to mentally and physically perform the task (i.e., to prevent or control the error and harm)? Are my evaluation skills good enough to propose, implement, and evaluate this prevention or control measure?

Like anywhere else, programs of prevention and control of medical error by improving knowledge, attitudes, or skills of individuals and groups require good operational definitions of these tools to improve the medical error problem and to make such interventions and their success or failure evaluable. Without it, improving the medical error problem by improving knowledge, attitudes, or skills may end up as hard-to-evaluate political statements and good intentions.

So far, some basic definitions are worth mentioning. Knowledge as part of the cognitive domain encompasses both retention of data and information about the subject and the capacity to apply them to specific tasks.[37] Improving cognitive functions to better manage the medical error problem also requires attention to other elements outlined by Bloom in the cognitive domain such as improving comprehension, application, analysis, synthesis, evaluation, and creativity.[38]

Attitudes as part of the affective domain may be seen as "learned tendencies to act in a consistent manner towards a particular object or situation," that is, to have the capacity to sense and recognize the situation, act in a controlled and predictable manner and to be consistent in a manner relevant to the situation. In relation to safety and health issues, these attributes need to be directly related to the situation and not of a generalist nature.[38,39] Attitudes may also be seen as part of Bloom's affective domain, including receiving, responding, valuing, organization, and characterization by a value or value complex.[38]

Skills as part of the psychomotor domain are defined as *expertness, practical ability, facility in doing something, dexterity,* and *tact.*[37] In this spirit, skills are part of the psychomotor domain including reflex movements, basic-fundamental movements, perceptual abilities, physical abilities, skilled movements, and non-discursive communication.[38] In summary, both reasoning and critical thinking problem solving and sensory-motor execution of tasks are skills.

In prevention and control health programs, knowledge, attitudes, and skills are considered independent variables that may lead to the alteration of medical error and harm problems.

The new paradigm for physicians-in-training now includes new knowledge, attitudes, and skills acquisition in their relationship with structure, process, and impact of activities in medical care. *See, do, teach* training in the patient safety domain should be expanded into a new training paradigm encompassing both knowledge/attitudes/skills and structure/process/impact of training and activities considerations. Rodriguez-Paz et al.[39] placed components of such training programs on patient safety in sequence from knowledge to attitudes, skills, and behavior to preventable harm.

Experimental, Quasi-Experimental, and Nonexperimental Evaluation of Interventions to Understand and Better Control Medical Error and Harm Problems

As already mentioned, the experimental approach to the study of causes of medical error is more than limited by most fundamental medical ethics (one cannot arbitrarily expose some subjects to medical error and some not and then study contrasts in error and harm frequency and severity between the groups) and by uniqueness or small number of cases.

In most instances, nonexperimental (observational or quasi-experimental) studies are the only way to evaluate if some preventive or control measures work.

Experimental methods as a controlled study of cause–effect relationships can be theoretically used in several situations:

- To confirm causes of error or harm.
- To demonstrate error as causing something else (like harm).
- To demonstrate harm as a cause of some additional outcome.
- To show that actions in prevention and control work; to show how effective they are in the improvement of error and harm situations.

Ethical reasons prevent or limit using experimental methods as a confirmation of the first three relationships in the human error and harm domains, even though such confirmation might be desirable and necessary in lathology.

Evaluation of the effectiveness of measures to improve error and harm events merits particular attention. However, its feasibility may be limited in the medical error domain. Clinical trials or controlled community interventions are not numerous.

In all instances in the health domain, we are interested in studies that are interventional and in the following properties:

- *Effectiveness*: how much good does the intervention produce in usual circumstances for a specific group of individuals like a proportion of events generated or prevented by the intervention.

- *Efficacy*: how the intervention works under ideal conditions.
- *Efficiency*: is the effect or end result of intervention proportionate to what was invested in it in terms of human and material resources and time.

These three types of information are complementary. Proper distinctions between them must be made when stating, even in lathology, that this or that program of error prevention works or not.

Hence, what is left in the methodological armamentarium to evaluate our actions and interventions in an effort to reduce the occurrence of medical error and limit the frequency, spectrum, and gradient of harm it produces?

Randomized or Otherwise Controlled Clinical Trials

Given the already mentioned complexity of modern interventions in the medical domain due to the number of interacting components within experimental and control interventions; to the behaviors required by those delivering or receiving intervention, groups, or organizational levels targeted by the intervention, number, and variability of outcomes; and to the degree of flexibility or tailoring of the intervention permitted, experimental designs for evaluating complex interventions include the following:[2]

- Individually randomized trials.
- Cluster randomized trials, stepped wedge designs.
- Preference trials and randomized consent designs.
- N-of-1 designs.

Despite some questions raised in the not too distant past,[40] the double-blind, randomized, placebo-controlled trial remains the reference method in the single evaluation of cause–effect relationships between medical interventions and improvement in health and disease. Systematic review and meta-analysis of multiple trials complete our view of the possible existence or nonexistence of some cause–effect relationship of interest, specifying in whom and in what circumstances. In the case of nonexistence, what are the alternatives? What is feasible in the medical error domain?

So far, experience with such an array of methodological approaches in lathology remains limited. The three examples that follow illustrate the role of randomized clinical trials in medical lathology.

In gynecology and obstetrics, clinical pharmacology, and pharmacy, Raebel et al.[41] from the Kaiser Permanente Colorado Clinical Research Unit conducted a randomized trial to determine whether a computerized tool that alerted pharmacists when a pregnant patient was erroneously prescribed contraindicated medications whose risk to the fetus outweighed therapeutic benefits (U.S. Food and Drug Administration [FDA] pregnancy risk category X) or for which there was evidence of fetal risk but therapeutic benefits could outweigh the risk (FDA category D) was effective in decreasing the dispensing of these medications. The control group involved women under usual care (no computerized alerting tool used). Figure 7.2 shows the flowchart of this trial.

This trial demonstrated that coupling data from information systems with knowledge and skills of physicians and pharmacists resulted in improved prescribing safety and lowered dispensing of erroneously prescribed drugs to pregnant patients.

In transfusion medicine, mistransfusion (transfusion of blood to the wrong patient) is considered to be the most important serious avoidable hazard (error) of transfusion. To prevent such a frequent bedside error, an international multicenter cluster-randomized, matched-paired clinical areas trial[42] showed that positioning a warning adhesive label ("STOP: Check the patient wristband") on the blood bag requiring the user to remove the label before attaching the administration set to transfuse red blood cells was not good enough and failed to improve bedside transfusion practice. There was no immediate or later favorable effect of this trial intervention.

Landrigan et al.[43] conducted a prospective randomized study to see if interns working in intensive care units according to a traditional schedule of tiring and stressful work shifts (24 hours and more) made more serious medical errors than those who avoided extended work shifts and had reduced the number of hours of work per week. The traditional schedule produced more serious medical errors, more nonintercepted serious errors, and more serious diagnostic errors. The authors concluded that the study's prospective and randomized nature allowed for a rigorous evaluation of the effects on patients' safety of an intervention designed to improve interns' sleep and thus decrease medical errors.[43] The harm to patients produced by errors was not evaluated in this study.

These three examples show that randomized trials are feasible in lathology wherever they are technically feasible and ethically acceptable. These studies also stress that both "positive" (effect seen) and "negative" (no effect) results are valuable in the assessment of the effectiveness of interventions to control medical error and harm.

Natural Experiment

Observation and analysis of possible causal associations may also be performed without experimenter intervention such as in allocations or other a priori methodological organizations of events. Subsets of the population that have different

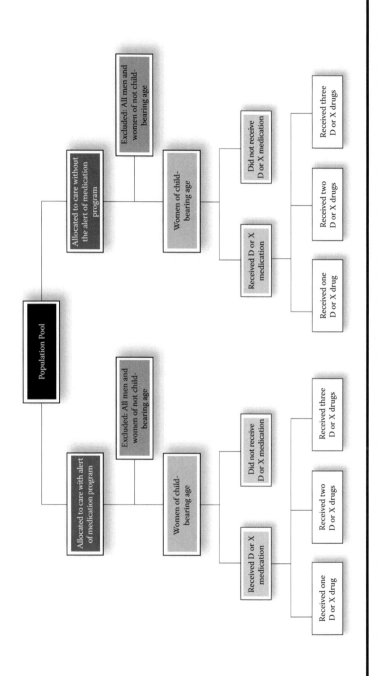

Figure 7.2 Trial in the medical error and harm domain. An example. (Redrawn with modifications from Raebel, M. A. et al., *J. Am. Med. Inform. Assoc.,* 14, 2007.[41])

levels of exposure to a supposed causal factor in a situation resembling an actual experiment, where human subjects would be randomly allocated to groups, are naturally occurring circumstances in which the presence of persons in a particular group is typically nonrandom. Yet for a natural experiment it suffices that their presence is independent of (unrelated to) potential confounders.[12]

Circumstances favorable to a natural experiment do not necessarily belong to classical infectious disease epidemiology. They might prove useful in detecting error in the medical error domain and analyzing and using them accordingly in medical error and harm explanation and control. Various follow-up studies (vide infra) might prove to be analyzable as natural experiments. Experience in lathology anyone?

Before–After Studies

Before–after studies represent, in our context, the follow-up of a health problem as it pertains to some crucial moment of implementation of an intervention expected to change the frequency or some other qualitative and quantitative characteristics of a target problem like a medical error and its consequences. As suggested in the title of such studies, at least two measurements of a given situation are made, one before and another after the implementation of the intervention of interest.

Before–after studies are usually based on one group of individuals: those exposed to the program. The effectiveness of implemented measures is based on the comparison of rates of events or other changes in their characteristics from one moment to another. For example, an international group of surgeons in eight nations/cities/hospitals implemented in 2007–2008 a Surgical Safety Checklist to be used before surgical interventions.[44] Anesthesia, surgery, and the performance of surgical nursing tasks are part of the checklist. The rate of postoperative complications (i.e., surgery-induced new morbidities), including death during hospitalization within the first 30 days after the operation, indicated the impact of this safety tool. The death rate declined from 1.5% before the checklist use to 0.8% afterward and inpatient complications fell from 11.0% before to 7.0% after.

In another before–after study, Espinosa and Nolan[45] compared rates of diagnostic errors in radiology made by emergency physicians before and after redesigning and implementing a system of interpreting radiographs (read by emergency physicians first; confirmations by radiologists later) with recalling of patients if important discrepancies were found. The rate of diagnostic errors and potential adverse effects diminished after the implementation of the new system. In yet other before–after studies, Potts et al.[46] noted an almost complete elimination of medical prescription errors and rule violations and a significant decrease of adverse drug events after the implementation of a computerized physician order entry system in a pediatric critical care unit. According to Rhodes et al.,[47] implementation of a national patient safety alert to reduce wrong-site surgery produced mixed results; it was effective in promoting presurgical marking and encouraging awareness of safety issues in relation to the correct site of surgery.

Readers may argue that before–after studies do not represent a more formal causal proof of medical error prevention and control by various measures implemented in various high-risk clinical working environments. They are right. However, various measures leading to various improvements and their heterogeneity across studies support to some degree the effectiveness of error control measures.

Reporting, surveillance, and time trends analysis of medical errors should provide additional information about the potential effects of various prevention and control measures that cannot be studied for numerous reasons in more convincing designs using, for example, control groups or controlled allocations to prevent bias in the medical error prevention programs' effectiveness.

Incident reporting itself is not enough to improve medical error situations.[48] It must be followed by further causal research and attempts to control error and harm incidents.

Within the framework of surveillance of medical error, which is often rare, *Monitoring Time to Next Medical Error*[49] was proposed as a way to assess the effect of changes in healthcare organizations and practice. Reducing frequency of medical errors and longer intervals between them should reflect the success of changing practices, activities, and care where errors and harm occur.

Case Studies

Case studies are detailed descriptions and analyses of the occurrence, development, and outcome of a particular problem or innovation, often over a period of time. They include a detailed description of a concrete situation requiring analysis, judgment, and sometimes action.[12] Case studies also provide at least workable hypotheses about medical error solving actions and interventions. New hypotheses must be as operational and explicit as possible to allow further research in intervention effectiveness. Although case studies may generate valuable information regarding the structure and process of approaches to medical error reduction,[50] they are not as powerful in the impact assessment of interventions.

Taking into account differences between errors and ensuing adverse events, case studies allow the detection of "triggers," or events that signal possible errors and harm, such as ordering of certain drugs, abnormal laboratory values, or abrupt stop orders. Trigger detection methodology[51] allows for a more complete detection and correction of medical error and harm and more timely intervention if possible.

Also, methodology combining case based reasoning and information retrieval from narratives included in event reports[52] helps identify clusters of similar events within clinical incident reports and focus on better interventions when feasible and required.

In another example,[53] case studies of errors made by medical students such as sterile techniques in the operating room, drugs prescribed versus drug administered, respecting the "do not resuscitate" orders or infection precautions complicated by language barriers led to proposals on how to improve student training in areas such as knowledge, attitudes, and skills learning including the knowledge of common causes and ways to control medical error.[53]

In any uses, medical errors are clinical events and are subject to clinical case reports whose content and form should be compatible with the methodology of general clinical case reporting as outlined elsewhere.[54] Clinical failures and unexpected disasters like mistakes in diagnosis or treatment and their consequences, unusual or unexpected effects of treatment (outcomes) or diagnostic results whether good or bad, diagnostic and therapeutic "accidents" (causes, consequences, remedies), unusual comorbidities (diagnosis, treatment, outcomes), unusual settings of medical care, uses of new technologies, complications of surgery (adverse effects, rejections, other failures, or intolerance of procedures), and tolerance of a procedure by the patient are among the reasons for clinical case studies, reporting, and publishing.[55]

Healthcare Failure Mode and Effect Analysis (HFMEA)

The Joint Commission of Accreditation of Healthcare Organizations (JCAHO) expects healthcare organizations to proactively identify high-risk processes, to predict and prioritize system weaknesses, to predict their outcomes, to determine why they occur, to adopt system changes to minimize the potential for patient harm, and to monitor the effectiveness of redesigned processes.[56]

In law, as pertains to health tort litigations (see Chapter 8), Green[57] feels that since not all errors can be prevented, it is necessary to reduce the consequence of the expression of error in the medical setting and that failure mode analysis is the appropriate tool to do so. This kind of analysis contributes to the identification of modes of medical errors, that is, particular appearances of the error if the error results in detectable and observable action, such as omissions, insertions, repetitions, or substitutions of error-generating and related activities.

Failure mode and effect analysis (FMEA) is defined as a procedure for the analysis of potential failure modes (descriptions of the way the failure occurs and the manner by which a failure is observed).[58] Initially developed by the military, it is now used in a variety of industries and in healthcare. It is a step-by-step approach for identifying all possible failures in organizational processes, policies, and procedures are reviewed, a design, a manufacturing or assembly process, or product or service.[59] In the health domain, FMEA is defined as a systematic, prospective, multidisciplinary team-based risk analysis that identifies and assesses the effects of potential errors or system failures.[56]

While root cause analysis as described in Chapter 5 represents a retrospective approach in the reconstruction of clinical events related to medical error, FMEA is rather prospective in following what happens and what might be drawn from ongoing experience to minimize failures and develop a new and better process of interest. In the domain of chemotherapy, prospective risk management potentially allows the prevention of errors rather than a reaction to them like in root cause analysis conducted after chemotherapy errors have occurred.

The FMEA concept was further adapted for the health domain by way of the HFMEA.[60] It is a five-step procedure:

1. The *topic* of the HFMEA is defined and justified as coming from a high-risk or highly vulnerability area and being a priority for analysis.
2. A multidisciplinary *team* including a subject matter expert, an advisor, and a team leader is assembled.
3. The *evaluation process* is graphically described in the form of a flow diagram listing the steps in the process and its subprocesses.
4. A *hazard analysis* follows. It is based on a list of well-defined failure modes for each of the subprocesses and on the assessment of their potential causes that are scored for their severity and probability leading to their prioritization for intervention and control. A Hazard Scoring Matrix is used for this purpose. A HFMEA decision tree is an algorithm (rather than a decision tree as understood and defined in the medical decision analysis domain) to determine which failure modes warrant further action.
5. A list of *actions and outcome measures* is developed including the description of actions for each selected failure mode of interest, what will be the outcome measures reflecting success or failure of corrective actions, and how they will be carried out, evaluated, and by whom.

HFMEA brings to lathology the expression *failure mode* designating ways a clinical or paraclinical activity and other type of care can fail to provide the anticipated result. This is a rather broad and encompassing concept including error, harm, and their determinants in the sense of their definitions reviewed in Chapters 2 and 3 of this book. For example, a failure mode may be not responding to an emergency call, mislabeling a medication, missing a needed instrument during an operation, or a breakdown in team communication.

The HFMEA Worksheet[60] is built in this spirit and contains several previously mentioned steps to understand and act in error and harm situations:

Step 1: Defining the HFME topic is essential from an epidemiological standpoint. It must be done in the best possible operational terms.

Step 2: Assembling the team requires bringing together individuals contributing to the evaluation process either by their expertise in the domain or by their methodological mastery of epidemiology, evaluation, and intervention.

Step 3: Graphically describing the process means the capture of essential elements of what happens in the activity that produces error and harm such as diagnostic clinical and paraclinical procedures or other elements of care with identification of possible and real failure modes. This is the descriptive stage of the HFMEA process.

Step 4: Hazard analysis, the understanding stage of HFMEA, starts by listing potential causes for identified failure modes. A two-step process follows:

Each potential cause is scored on a directional category scale or quantitatively like from 1 to 10 first[61] *for its severity* (catastrophic, major, moderate, or minor severity of effect degree probability) and for its probability (frequent, occasional, uncommon, or remote) *to be combined into a hazard score. Detectability is also scored* this way. In the nonmedical world,[62] the FMEA hazard analysis step may be based on a 1–10 grading of occurrence, severity, and chance of detection giving a combined score for risk priority for intervention and control.

A *decision tree analysis* in the HFMEA context means here rather an algorithm based on the yes or no considerations of likelihood of severity and occurrence, presence of a single point of weakness, existence of an effective control measure, and obviousness of the hazard (detectability). It is a direction-giving tool in which each step depends on the previous one rather than a decision tree elsewhere in medicine, which is a direction-seeking one based on various options as is understood in the medical decision-making domain.

Step 5: Actions and outcome measures are developed, described, implemented, and evaluated. The fifth step of the HFMEA is then the *action or intervention development and decision stage.* An action type is selected, the rationale for stopping it is specified, and outcome measures are chosen. Administrative information is added like who will be responsible for action implementation or evaluation of aspects like monitoring the effect of corrective measures and redesigning this step according to the monitoring results.

HMFEA requires some a priori knowledge of possible frequency and causes of events of interest and rather subjective attribution of probabilities related to them.

Good FMEA severity, occurrence, and detection definitions as the basis for the hazard degree assessment in general, beyond medicine (hazard scores), were proposed for iSix Sigma by Curtis.[61]

A particular HFMEA application comes from the chemotherapy domain.[56] Once a FMEA (HFMEA) team is constituted, organizational processes, policies, and procedures that apply to chemotherapy or any other healthcare intervention are reviewed. Actual practice conflicts with written policies and procedures lead to the further defining of potential failure modes. Failure modes include everything that could go wrong that would prevent the subprocess step within the whole process from being carried out.[63]

Once they are defined, the potential effect of each failure on patients is determined, the probability, severity, and detectability of each failure is ranked (hazard scores here are the product of severity and probability scores of various failure modes) and the areas of greatest concern emerge from such an evaluation. The possibility of failure is reduced by determining what would cause a step in the process to fail. (Root cause analysis may be attempted for this purpose.) The FMEA team develops a new process and establishes a plan to measure success. The new plan is implemented if its superiority is supported by a pilot plan that precedes it. The success of the new plan is evaluated.[56]

In traumatology, the Healthcare FMEA was used by Day et al.[64] to identify strategies that would reduce risks and improve patient safety during registration of trauma patients and subsequent electronic data linkage.

Weaknesses in the system of trauma patient registration were identified, leading to the implementation of changes that included education of staff, role clarification, task reallocation, and establishment of a list of personnel authorized to request electronic data linkage process.

Failure mode and effect analysis were also proposed for identifying risk in community pharmacies using a modified eight-step FMEA process:

1. Selecting the process and assembling the team.
2. Diagramming the process.
3. Brainstorming potential failure modes and determining their effects.
4. Identifying the causes of failure modes.
5. Prioritizing failure modes.
6. Redesigning the process.
7. Analyzing and testing changes.
8. Implementing and monitoring the redesigned process.[65]

This process follows more closely the general health problem evaluation path from its definition to its execution and evaluation of its impact.

Looking at healthcare as a process and service producing desirable health outcomes, FMEA (HFMEA) may be used (in the language of manufacturing and industry) among others:

- When a process, product, or service is being designed or redesigned.
- When applied in a new way.
- When improvement goals are planned for an existing process, product, or service.
- When analyzing failures of an existing process, product, or service (ASQ).

Like root cause analysis (RCA), the HFMEA operates on the basis of the study of factors whose causal role is already presupposed, demonstrated, proven, or anticipated on the basis of already acquired experience. The demonstration and evaluation of newly discovered causal factors belongs to formal etiological research methodology.

Systematic Reviews of Evidence

The desired and fortunate increase in the number of clinical trials and other studies of medical error prevention effectiveness begins to lead to equally welcome and needed systematic reviews and meta-analyses. For example, Mhyre et al.[66] systematically reviewed randomized controlled trials that evaluated strategies to avoid epidural vein cannulation during obstetric epidural catheter placement. Various actions were covered such as patient position, anatomic approach, needle and catheter gauge, fluid injection before catheter insertion, multiforce catheters, wire-embedded polyurethane catheter design, and epidural catheter insertion depth. Among those seven considerations of actions that may lead to error in obstetric epidural catheter placement, their review shows that a reduction in vein cannulation was achieved by positioning the patient in a lateral as opposed to a sitting posture, injecting fluid through the epidural needle before the catheter insertion, using a single catheter instead of a multi-orifice one, using a wire-embedded polyurethane catheter instead of a nylon one, and limiting the depth of catheter insertion to 6 cm or less.

Completeness and quality of evidence was not omitted in the authors' recommendations of higher quality studies to test the interaction between the various actions and strategies previously mentioned.

Conclusions and Recommendations

Across current experience in the medical error control domain, many valuable contributions were made to solve this problem, and its specific methodology is becoming increasingly rich. In the meantime, we often forget that medical error or harm are, in the eyes of a clinical epidemiologist, another special case of a health event (like disease) that might benefit from epidemiological approaches to its occurrence, causes, and preventive and otherwise corrective interventions and their evaluation.

Several points will direct our initiatives and priorities in the future:

- As we see in this overview of actions and their evaluation in the medical error domain, corrective measures focus mainly on flaws in the system rather than on flaws in human individual understanding and decision making. Individual actions are evaluated as part of the "system" rather than as entities of interest by themselves.
- Uses of the experimental method of intervention or evaluation and their epidemiological interpretation are limited by the low frequency in error occurrence and the fact that errors lead to harm. It is not ethical to deliberately lead humans to exposure to harmful effects.
- Current experience with experimentation, corrective intervention, and their evaluation has been directed so far mainly toward structural and procedural changes rather than toward the qualitative and quantitative evaluation of results in epidemiological terms. It's done often in the hope that structural and procedural changes will lead to changes in impact, which is not necessarily so in all cases.
- Transfer of experience from nonmedical domains like manufacturing, industry, military arts, operational research, or business and other domains creates vocabularies, terminologies, and taxonomies that are not necessarily uniformly understood and used across the current experience in the medical error and harm domain. However, adaptations and uses of the experience in the nonmedical domain are increasingly beneficial in medicine and other health sciences domains. Nursing remains particularly active in such initiatives.
- The still persisting interchangeable use of *error* and *harm* creates subjects and their studies that may be methodologically sound in the error domain but not in the harm domain and vice versa. Overlaps may exist.
- It is still easier to evaluate the structure or process of activities in clinical and community medical care than its impact.
- Modifications of interventions in human knowledge, attitudes, and skills and their evaluation as modifiers of the medical error experience must still be further evaluated and understood. Most of them belong to the category of "soft" data, which are harder to observe, measure, define, and otherwise manipulate than "hard" data like morbidity, mortality, or case fatality rates so valued by most mainstream epidemiologists.

Current intervention experience in the medical error domain is richer in error and harm protection and primary prevention than in secondary and tertiary prevention and promotion of error and harm-free human and material environments. A better balance in the future may prove of great value in our desire to control the medical error problem.

It would be unfounded to say that many medical error problems cannot be solved by a formal epidemiological approach, whatever the reason might be. Complementary methodological epidemiological and nonepidemiological approaches in the future are definitely worthy of interest. Randomized, multiple-blind, multiple-controlled clinical or field (community medicine and primary care) trials may be ideal in current causal proofs of the effectiveness and efficacy of interventions, but they cannot be performed everywhere and applied to any problem to be solved, however much we would like it to be so from an epidemiological perspective.

An increasing number of original experiences will certainly lead to an increasing number of systematic reviews and meta-analyses that will further rationalize lathology in the intervention domain.

Many good steps have been taken so far, and the future looks even more promising.

References

1. Pronovost PJ, Colantuoni E. Measuring preventable harm. Helping science keep pace with policy. *JAMA*, 2009;**301**(12):1273–5.
2. Craig P, Dieppe P, Macintyre S, Mitchie S, Nazareth I, Petticrew M. Developing and evaluating complex interventions: the new Medical Research Council guidance. *BMJ*, 2008;**337**(Oct 25), 979–83.
3. Nolan TW. System changes to improve patient safety. *BMJ*, 2000;**320**(March 18), 771–3.
4. Battles JB, Lilford RJ. Organizing patient safety research to identify risks and hazards. *Qual Saf Health Care*, 2003;**12**(Suppl II): ii2–ii7.
5. Volpp KGM, Grande D. Resident's suggestions for reducing errors in teaching hospitals. *N Engl J Med*, 2003;**348**(Feb 27):851–5.
6. Brown C, Hofer T, Johal A, Thomson R, Nicholl J, Franklin BD, et al. An epistemology of patient safety research: A framework for study design and interpretation. A series of articles in *Qual Saf Health Care*, 2008;**17**:158–81.
7. Leape LL. Scope of problem and history of patient safety. *Obstet Gynecol Clin N Am*, 2008;**35**:1–10.
8. Oyebode F. Clinical errors and medical negligence. *Adv Psychiatr Treat*, 2006;**12**:221–227.
9. Keohane CA, Bates DW. Medication safety. *Obstet Gynecol Clin N Am*, 2008;**35**:37–52.
10. Cusack CM. Electronic health records and electronic prescribing: promise and pitfalls. *Obstet Gynecol Clin N Am*, 2008;**35**:63–79.
11. Nielsen P, Mann S. Team function in obstetrics to reduce errors and improve outcomes. *Obstet Gynecol Clin N Am*, 2008;**35**:81–95.

12. *A Dictionary of Epidemiology.* Fifth Edition. Edited by M Porta. New York: Oxford University Press, 2009.
13. Leape LL, Woods DD, Hatlie MJ, Kizer KW, Schroeder SA, Lundberg GD. Promoting patient safety by preventing medical error. *JAMA,* 1998;**280**(16):1444–7.
14. Vincent C, Taylor-Adams S, Stanhope N. Framework for analysing risk and safety in clinical medicine. *BMJ,* 1998;316:1154–7.
15. Cook RI, Woods DD. Operating at the sharp end: the complexity of human error. Pp. 255–310 in: *Human Error in Medicine.* Edited by S Bogner. Hillsdale, NJ: Erlbaum, 1994.
16. Nolan TW. System changes to improve patient safety. *BMJ,* 2000;**320**(March 18): 771–3.
17. Bates DW. Preventing medical errors: A summary. *Am J Health-Syst Pharm,* 2007;**64**(Suppl 9, Jul 15):53–9.
18. Frush KS. Fundamentals of patient safety program. *Pediatr Radiol,* 2008;**38**(Suppl 4):S685–S689.
19. Dute J. Recommendation Rec(2006)7 of the Committee of Ministers to Member States on Management of Patient Safety and Prevention of Adverse Events in Health Care. *Eur J Health Law,* 2008;**15**:79–98.
20. Etchells E, O'Neill C, Bernstein M. Patient safety in surgery: error detection and prevention. *World J Surg,* 2003;**27**(8):936–42.
21. Bohnen JMA, Lingard L. Error and surgery: Can we do better? *Can J Surg,* 2003;**46**(5):327–9.
22. Wikipedia, the Free Encyclopedia. *Preventable medical error.* Available at: http://mhtml:file://K:/Preventable medical error – Wikipedia, the free encyclopedia.mht, retrieved April 10, 2009.
23. *A Dictionary of Epidemiology.* Fourth edition. Edited by JM Last. Oxford: Oxford University Press, 2001.
24. Weiss CH. *Evaluation Methods for Studying Programs and Policies.* 2nd Edition. Upper Saddle River, NJ: Prentice-Hall, 1988.
25. An overview of program evaluation. Pp. 1–30 in: Rossi PH, Lipsey MW, Freeman HE. *Evaluation. A Systematic Approach.* 7th Edition. Thousand Oaks, CA: Sage Publications, 2004.
26. Jenicek M. *Foundations of Evidence-Based Medicine.* Boca Raton, FL: Parthenon Publishing Group/CRC Press, 2003.
27. *Wallace/Maxcy-Rosenau-Last Public Health & Preventive Medicine.* Edited by RB Wallace, N Kohatsu, JM Last. New York: McGraw Hill Medical, 2008.
28. *Oxford Textbook of Public Health.* Third Edition. Edited by R Detels, WW Holland, J McEwen, GS Omenn. New York: Oxford University Press, 1997.
29. Donabedian A. Evaluating the quality of medical care. *Milbank Mem Fund Quart,* 1966;**44**(3):166–206.
30. Basic approaches to assessment: Structure, process, and outcome. Pp. 77–128 in: *Explorations in Quality Assessment and Monitoring. Volume I. The Definition of Quality and Approaches to Its Assessment.* Edited by A Donabedian. Ann Arbor, MI: Health Administration Press, 1980.
31. Fink A. *Evaluation Fundamentals. Insights into the Outcomes, Effectiveness, and Quality of Health Programs.* Thousand Oaks, CA: Sage Publications, 2005.

32. Kunkel S, Rosenquist U, Westerling R. Implementation strategies influence structure, process and outcome of quality systems: an empirical study of hospital departments in Sweden. *Qual Saf Health Care,* 2009;**18**:49–54.

33. The justification of surgery. Pp. 95–162 in: *Explorations in Quality Assessment and Monitoring. Volume III. The Methods and Findings of Quality Assessment and Monitoring. An Illustrated Analysis.* Edited by A Donabedian. Ann Arbor, MI: Health Administration Press, 1985.

34. Bloom BS, Krathwohl D. T*axonomy of Educational Objectives: The Classification of Educational Goals. Handbook I. Cognitive Domain.* New York: Longmans, Green 1956.

35. Krathwohl DR, Bloom BS, Bertram MS. *Taxonomy of Educational Objectives: The Classification of Educational Goals. Handbook II. Affective Domain.* New York: David McKay Co.,1964.

36. Bloom BS. *Taxonomy of Educational Objectives.* Boston: Allyn and Bacon, 1984.

37. Anon. *What are knowledge, skills and attitudes?* Available at: http://www.apaseq.com/docs/knowledge.doc, retrieved May 2, 2009.

38. Anon. *The taxonomy of educational objectives.* Available at: http://www.humboldt.edu/~tha1/bloomtax.html, retrieved May 17, 2008.

39. Rodriguez-Paz JM, Kennedy M, Wu AW, Sexton JB, Hunt EA, Pronovost PJ. Beyond "see one, do one, teach one:" toward a different training program. *Qual Saf Health Care,* 2009;**18**:63–8.

40. Kaptchuk TJ. The double-blind, randomized, placebo-controlled trial: Gold standard or golden calf? *J Clin Epidemiol,* 2001;**54**(6):541–9.

41. Raebel MA, Carroll NM, Kelleher JA, Chester EA, Berga S, Magid DJ. Randomized trial to improve safety during pregnancy. *J Am Med Inform Assoc,* 2007;**14**:440–50.

42. Murphy MF, Casbard AC, Ballard S, Shulman IA, Heddle N, Aubuchon JP, et al. Prevention of bedside errors in transfusion medicine (PROBE-TM) study: a cluster-randomized, match-paired clinical areas trial of simple intervention to reduce errors in the pretransfusion bedside check. *Transfusion,* 2007;**47**(May):771–80.

43. Landrigan CP, Rothschild JM, Cronin JW, Kaushal R, Burdick E, Katz JT, et al. Effect of reducing interns' work hours on serious medical errors in intensive care units. *N Engl J Med,* 2004(Oct 28);**351**(18):1838–48.

44. Haynes AB, Weiser TG, Berry WR, Lipsitz SR, Breizat A-HS, Dellinger EP, et al. A surgical safety checklist to reduce morbidity and mortality in a global population. *N Engl J Med,* 2009;**360**;5(Jan 29):491–9.

45. Espinosa JA, Nolan TW. Reducing errors made by emergency physicians in interpreting radiographs: longitudinal study. *BMJ,* 2000(March18);**320**:737–40.

46. Potts AL, Barr FE, Gregory DF, Wright L, Patel NR. Computerized physician order entry and medication errors in a pediatric critical care unit. *Pediatrics,* 2004;**113**(1):59–63.

47. Rhodes P, Giles SJ, Cook GA, Grange A, Hayton R, Maxwell MJ, et al. Assessment of the implementation of a national patient safety alert to reduce wrong site surgery. *Qual Saf Health Care,* 2008;**17**:409–15.

48. Vincent C. Incident reporting and patient safety. Emphasis is needed on measurement and safety improvement programmes. *BMJ,* 2007;**334**(Jan 13):51.

49. Hovor C, Walsh C. Tutorial on monitoring time to next medication error. *Q Manage Health Care,* **16**(4):321–7.

50. Nicol N. Case study: An interdisciplinary approach to medical error reduction. *Am J Health-Syst Pharm,* 2007;**64**(Suppl 9): S17–S20.

51. Resar RK, Rozich JD, Classen D. Methodology and rationale for the measurement of harm with trigger tools. *Qual Saf Health Care,* 2003;**12**(Suppl II):ii39–ii45.

52. Tatsoulis C, Amthauer HA. Finding clusters of similar events within clinical incident reports: a novel methodology combining case based reasoning and information retrieval. *Qual Saf Health Care,* 2003;**12**(Suppl II):ii24–32.

53. Seiden SC, Galvan C, Lamm R. Role of medical students in preventing patient harm and enhancing patient safety. *Qual Saf Health Care,* 2006;**15**:272–6.

54. Jenicek M. *Clinical Case Reporting in Evidence-based Medicine.* Second Edition. London: Oxford University Press, 2001.

55. Jenicek M. Clinical case reports and case series research in evaluating surgery. Part II. The content and form: Uses of single clinical case reports and case series research in surgical specialties. *Med Sci Monit,* 2008;**14**(10):RA149–RA162.

56. Sheridan-Leos N, Schulmeister L, Hartfant S. Failure Mode and Effect Analysis™: A technique to prevent chemotherapy errors. *Clin J Oncol Nurs,* 2006;**10**(3):393–401.

57. Green M. *Human Error in Medicine.* Available at: http://www.expertlaw.com/library/malpractice/medical_error.html, retrieved March 26, 2009.

58. Wikipedia, the Free Encyclopedia. *Failure mode and effect analysis.* Available at: http://en.wikipedia.org/wiki/Failure_mode_and_effects_analysis, retrieved February 3, 2009.

59. American Society for Quality (ASQ). *Failure Modes and Effect Analysis (FMEA).* Available at: http://www,asq,org/learn-about-quality/process-analysis-tools/overview/fmea.html, retrieved February 3, 2009.

60. DeRosier J, Stalhandske E, Bagian JP, Nudell T. Using Health Care Failure Mode and Effect Analysis™: The VA National Center for Patient Safety's Prospective Risk Analysis System. *J Qual Improv,* 2002; **28**(5):248–67.

61. Curtis S. *FMEA Severity, Occurrence, and Detection Definitions.* Available at: http://main.isixsigma.com/forum/showmessage.asp?messageID=15024, retrieved May 9, 2009.

62. Concordia University. *Failure mode and effect analysis. What it is: When to use it: How to use it.* Available at: http://web2.concordia.ca/Quality/tools/11failuremodeanalysis.pdf, retrieved May 9, 2009.

63. Stanhandske E, DeRosier J, Patail B, Gosbee J. How to make the most of failure mode and effect analysis. *Biomed Instr & Technol,* 2003; **37**(2 -March/April):96–102.

64. Day S, Dalto J, Fox J, Allen A, Ilstrup S. Utilization of failure mode effects analysis in trauma patient registration. *Qual Manag Health Care,* 2007;**16**(4):342–8.

65. Greenall J, Walsh D, Wichman K. Failure mode and effect analysis: A tool for identifying risk in community pharmacies. *CPJ/RPC,* 2007(May/June);**140**(3):191–3.

66. Mhyre JM, Greenfield MLVH, Tsen LC, Polley LS. A systematic review of randomized controlled trials that evaluate strategies to avoid epidural vein cannulation during obstetric epidural catheter placement. *Obstet Anesthesiol,* 2009;**108**(4):1232–42.

Chapter 8

Taking Medical Error and Harm to Court: Contributions and Expectations of Physicians in Tort Litigation and Legal Decision Making

Executive Summary

Medical errors and the harm they potentially produce are frequent subjects of court litigation. Physicians and other health professionals play two major roles in courts of law. First, they are held responsible for harm to the patient or community that resulted from their professional errors in breach of standard care; they then may and can become defendants in malpractice matters being sued by their patients. Second, they are called to courts by any involved party in litigation as experts in health matters to enlighten the judge, lawyers, and juries about the nature, causes, magnitude, management, and prevention of medical errors and harm. In any of these roles, physicians and other health professionals must be ready to answer questions about the nature of the error and harm, the magnitude of its impact on the patient, the causes of medical error, and if the medical error was a cause of the patient's harm.

In terms of modern argumentative exchanges of ideas, statements at courts must be seen as claims or conclusions resulting from argumentation. Any party at courts—be it plaintiffs, defendants, or witnesses—must be prepared to provide courts not only with all evidence required for the judicial process and judgment but also with all necessary premises, grounds, backing, warrants, qualifiers, and rebuttals supporting their statements. Strategies in philosophy and law may differ.

Demonstrations of general and case-specific cause–effect relationships in medical error analysis rely heavily on the clinical and field epidemiology methodology, data, and information produced. Incomplete and not necessarily representative epidemiological information about the medical error and harm limits the causal proofs and remedial and preventive measures. Frequently, a disclosure challenge lies behind the quality and complexity of epidemiological information available.

Health and law professionals must first agree on often widely different meanings of terms and their uses in the field of medical errors. Once an agreement has been reached, they must be ready to be involved, if necessary, at all stages of the law process or lawsuit: before the trial, at the trial, and after the trial. Generally, the burden of proof rests with the plaintiff. The defendant's response in the case of an alleged medical error, the jury's views, and the judge's conclusions and decisions all rely heavily on the physician and other health professionals on the stand (i.e., expert evidence and defendant's evidence). The common language of medicine and law and mutual understanding are among the cornerstones of litigation at courts of law.

Thoughts to Think About

If a doctor has treated a man with a metal knife for a severe wound, and has caused the man to die, his hands shall be cut out.

—Hammurabi's Code, ~2000 BC

There's a sort of decency among the dead, a remarkable discretion; you never find them making any complaint against the doctor who killed them.

—Molière, 1622–1673

And it should be the duty of physicians and surgeons to defend their art before the Courts and not permit further abuse of this art which is of so great importance.

—Pierre Franco, 1500–1561

Mala praxis: Injuries by the neglect or unskilful management of a person's physician, surgeon and apothecary, breaks the trust which the party had placed in his physician, and tends to the patient's destruction.

—Sir William Blackstone, 1768

There can be no malpractice without established practice; a physician cannot be convicted of deviating from the accepted standards if no accepted standards exist.

—James C. Mohr, 1943–

We've all seen Apollo 13. NASA guys always have the old "plastic bag, cardboard tubing, and duct tape" option to fall back on when the shit hits the fan. Neurosurgeons have not such a leeway. What it comes down to is this: if a brain surgeon screws up, it means a multi-million-dollar malpractice suit, but if a rocket scientist screws up, it means a multi-million-dollar hit movie starring Tom Hanks.

—Michael J. Fox, 2002

A physician does not end up at courts because he or she has committed a medical error, but rather because of the harm imputable to a medical error. The connection to error, imputability of harm, and its compensation lies with courts of law and their analysis and decisions.

Individual patients and entire communities sometimes complain and sue when they doubt medical decisions, the effects of industrial and other environmental factors, or their possible role in perceived poor health. Did a physician's care and decisions during a difficult labor lead to a newborn's brain damage? Was it human error or a system error or both? What caused the error? Did the error lead to harm? Did exposure of a population to air pollutants in an urban area produce a high occurrence of respiratory problems? Were these problems due to some kind of industrial or community policy error or something else? Did toxic waste disposal lead to leukemia outbreaks and clusters? Was cancer in military veterans due to exposure to chemical warfare agents? Did a physician make an error in his or her practice related to some of these? Only recently, with more attention given to the analysis and assessment of the role of error itself in these matters, has experience paid off.

We addressed in part some of the challenges of medicine and epidemiology at courts of law in Foundations of Evidence-Based Medicine.[1] In the spirit of that discussion, what should be added and expanded in view of medical errors and harm? Let us review in this chapter what has already been said in a broader context and what is particularly relevant in medical, clinical, and community medicine/public health lathology.

Introductory Comments

Politicians are becoming involved in making patient safety the centerpiece of some medical liability reform. In the United States, the *National Medical Error Disclosure and Compensation (MEDiC) Program,* as outlined in 2006, is a good example of federal initiatives, proposed by two lawyers: Hillary Rodham Clinton and Barack Obama.[2] Those and other strategies focus on system error rather than on human error. Not only the improvement of failed systems and procedures (and not the negligence of physicians) but also more effective communication between patients and providers should help in some way control liability and tort litigations costs. Their legislation[2] [(HRC, BO), The National Medical Error Disclosure and Compensation (MEDiC) Bill (S. 1784)], was introduced; the system of disclosure of "medical errors" also encompasses system errors. Horton feels, however, that the debate should start between doctors.[3] In radiology, Olivetti et al.[4] believe that a better understanding and management of medical error should first focus on errors in perception, reasoning (e.g., diagnostic errors, hindsight challenge, ensuing likelihoods, clinical trade-off between false positives and false negatives), and decision making of radiologists leading to errors both of commission and omission. System considerations follow, such as available technology, setting, work hours, or time dedicated to reporting. Foreseeability, avoidability, certainty of errors, and the possible ability to rectify the misconduct evaluation should also be part of medical liability assessment.[4]

Liang[5] proposes that legal reforms covering medical error and liability should be based on a better analysis and understanding of the errors produced by health system investigatory programs inspired by the experience of error management in the aviation or nuclear power industry and that health professionals should be more like partners in such activities than subjects to tort litigation.[5]

Gilmour[6] presents an excellent comparative study of patient safety, medical error, and tort law contrasting medical error and its legal management and control strategies in Canada, the United States, the United Kingdom, Australia, and New Zealand.[6] Her vision and emphasis is on the system, or "blunt-end," characteristics of error in the health domain rather than on the "sharp-end" errors made by individuals. Ongoing reforms of the medical liability system must take into account system characteristics and challenges like medical error disclosures and reporting and their monitoring and evaluation while benefiting from the sharing of experience from lawsuits and implementing compensation for physicians, hospitals, and their patients (including government contributions).

But first and foremost, physicians and health professionals should be more familiar with what happens in the legal system, how they will (in reality, most will not) become involved in this area including in medical tort litigation, how to understand its basics, and what their expected role is either as defendants or expert witnesses. This chapter covers some of the essential points to ponder from a medical perspective. Readers should be aware of national and international differences in the definition, perception, application, and uses of the medical error and harm problem and will certainly adjust our more general message to their specific conditions and applications.[6-8]

The notion of "evidence" is important not only for physicians in times of "evidence-based medicine" but equally, if not more broadly, in the field of law. A common language is currently being constructed,[9,10] and common understanding is vital wherever and whenever medical findings are used "in the real world," such as in tort litigation in courts of law. Physicians and other providers of health information who may be potential expert witnesses (but also defending themselves) should all know the fundamentals of their legal obligations and how they may be enforced.

Physicians play two major roles in the legal world:

■ Physicians and other health professionals are brought to courts to assess if they are (or not) responsible for some health (injury) or monetary damage to the plaintiff, be it the patient or his or her family, community, employer, or insurance company. Very often, the question of whether they are guilty of malpractice arises.
■ Physicians and epidemiologists are increasingly called upon as expert witnesses to express their opinions about types of cause–effect relationships and standard of care and are asked to comment on who else is at fault. Ultimately, the court determines whether a party is at fault.

Some of the most frequent questions in the area of error and harm that are asked of physicians are as follows:

■ Did some error occur? What kind of error is it? They may also be asked to define it.
■ What caused that error?
■ What was the outcome of that error? Did it lead to harm? What kind of harm?
■ What is the standard of care in the circumstances?

Not all errors breach the standard of care. These types of questions may be at the core of some malpractice litigations.

All health professionals (not only epidemiologists) must be cognizant of the rigor of causal proof in law and its differences from causal proof in medicine. Considerable amounts of money and the health and well-being of subjects are at stake. Conclusions are made based on solid evidence and not on hearsay or demagogic statements from positions of authority, function, or qualification. Meta-analysis[11,12] will play an increasingly important role in such endeavors. Causal demonstrations in courts are challenging and difficult; they may differ from such demonstrations in medicine and in some cases can lead to different results. Credibility, competence, and experience of all experts involved must be rebuilt from scratch before the judge. Everything is recorded, and records become public domain.

Judges and lawyers are now well aware of epidemiology, and the issue of causal proof in health and disease is the subject of attention in law and medicine[13-20] as well as in a growing body of literature. Surprisingly for many, the reasoning processes of people of law and medicine are often very similar, sometimes not. "Logic for lawyers"[21] and "logic for physicians"[22] are closely interwoven, but the burden of proof is different in law than it is in medicine, although results may be the same in the end.

Both medical error and harm are the subject of cause–effect considerations just as disease or any health phenomenon are causes and consequences in medicine and all other health sciences. Let us consider the following example:

Most recently, epidemiological experience from the investigation of causes of outbreaks was applied to such unusual "outbreaks" as a series of clinical course complications or patient deaths in hospitals that might be related to such a cause as a mentally disturbed or criminal employee, or any other responsible "cause" of such a situation. What kind of error was committed, and what kind of harm followed? Multiple cause–effect hypotheses (which employee might be related to the occurrence of cases) must be and have been successfully analyzed in courts as contributing evidence to strengthen final conclusions about the "guilt," (i.e., liability of error committed by suspected individuals).[23-26] Those strategies may be used in the study of error that might be potentially responsible for this kind of harm. But cause–effect considerations are more general in the medical/legal domains.

Analyzing medical error, its causes, and consequences through epidemiological eyes is currently based on data that are reported and recorded with a various degree of complexity and completeness. Fundamental epidemiological information about the error and harm occurrence is still missing. These events may be declared to some degree, but there is no system of compulsory declaration and registry of "notifiable cases of error and harm" as we have for compulsory reporting of "notifiable infectious diseases." Hence, any potential and existing surveillance system allowing epidemiological forecasting and rapid intervention in case of need is more than limited. The incidence rate of error and harm is hard, if not impossible, to establish; denominators for incidence rates remain obscure, and the notion of incidence density (persons/time at risk of error and harm) is generally unused and unexplored. Analyses of causal associations and their proof in the error and harm domain have a value dependent only on the epidemiological material, data, and information defined by what was recorded in terms of completeness, knowledge of population, time setting, and information gathering methodology. "General" causal associations obtained are scrutinized to see if they apply to the individual; in law, focus tends to be on the individual plaintiff or defendant before the court. This problem is inherent to any medico-legal discussion of error and harm.

In many cases, courts are referred only to single cases or small case series as a basis for reference on causality and ultimately for judicial decisions and verdicts.

Medical, Surgical, and Public Health Malpractice Claims and Litigation

How do medical errors fit into the current state of legal dealings with what is historically called malpractice? Are medical errors synonymous with malpractice? Will they automatically end up in courts of law?

Medical and Surgical Malpractice

For malpractice in the domain of clinical care, let us retain the following definition: "Any professional misconduct, unreasonable lack of skill or fidelity in professional duties or illegal or immoral conduct.... In medical practice, nursing practice, and allied professions, malpractice means bad, wrong, or injudicious treatment of a patient. Professionally, it results in injury, unnecessary suffering, or death to the patient." (The court may hold that malpractice has occurred even though the practitioner acted in good faith.[27])

Mohr, in his historical review of medical malpractice litigation in the United States, reminds us that as early as 1768, Briton Sir William Blackstone was raising the question of *mala praxis,* concerned that "injuries by the neglect or unskilful management of a person's physician, surgeon and apothecary, breaks the trust which the party had placed in his physician, and tends to the patient's destruction."[28] The growth in the number of medical malpractice lawsuits is due not only to legal factors; it is also paradoxically attributed to positive events like the development and uses of new medical technologies, the spread of uniform standards, the availability of medical malpractice liability insurance, and the growing ability to keep insured persons alive, therefore requiring costly treatment and supports. Either way, malpractice litigation may appear as barriers to quality improvements because they lead to the defensive practice of medicine, error secrecy, and underreporting of events.[29]

Medical errors play a variable role in malpractice claims. In the United States, Studdert et al. analyzed 1,452 claims, of which 37% did not involve errors. A total of 73% of claims that involved injuries due to error were compensated. The most frequently sued specialties were, in decreasing order, obstetrics-gynecology, general surgery, and primary care.[30] In surgery, prevailing types of error found in malpractice claims were errors of execution (i.e., slips and lapses in lathology terms),[31] manual errors were found in 65% of cases, and 9% were related to failure in knowledge and judgment (i.e., mistakes).[31] Handoffs figured among the system errors involved. Technical errors were involved most often in routine operations conducted by experienced surgeons dealing with complex patients and circumstances.[32,33]

Rogers et al.[34] found among factors contributing to malpractice claimed injuries attributable to surgical errors: errors in judgment in 66% of cases of malpractice claims; failure of vigilance/memory in 63% of cases; lack of technical competence or knowledge in 41% of cases; and communication breakdowns in 24% of cases.

Making medical mistakes is not necessarily "malpractice."[35] It must be demonstrated as such.

Public Health Malpractice

Public health malpractice still awaits its own definition. Essentially, it should reflect (in the generation and management of health error and harm) additional involvement of individuals and bodies other than health professionals: civic administration and governments and other stakeholders in health beyond hospitals, clinics, and various health units manned by health professionals, administration, and supporting services. Disease affecting groups and communities beyond individual cases and its consequences often becomes a unit of observation and interest. It has its own web of causes and consequences as a mass phenomenon in addition to ways of controlling it.

Public health malpractice differs from malpractice in clinical care. Two types of malpractice may occur in public health for which both health professionals and civic bodies and nonhealth professionals in industry, manufacturing, transportation, or communications may also be held responsible.

Two parties may be involved:

■ First, *public health malpractice* may be committed *by health professionals*. Epidemiologists, medical officers of health, and other community medicine specialists may commit errors in their management of public health problems like failing to investigate and control outbreaks or pandemics and omitting risks in occupational medicine or infectious and noninfectious disease surveillance or forecasting. Such errors may be of a medical nature and fall under the category of medical errors as mentioned in the previous section. They are usually handled as medical errors. However, they are found in courts of law less often than clinical practice-generated medical errors.

■ Second, *public health malpractice* may also be committed *by nonhealth professionals,* administrators, and various civic bodies and organizations responsible for healthy living conditions, lifestyle, and protection against noxious agents. Air-, water-, and soil-polluting industries, industrial uses, and exploitation of ecosystems or development and implementation of health risk-generating new technologies are terrains for public health malpractice. The responsibility for them is complex, and its legal management lies well beyond the scope of medical problems as discussed in this book.

In both cases, class actions may be possible because of more than one individual claiming harm; also, there may be multiple plaintiffs in a nonclass suit. In both cases, physicians may become defendants presumably responsible for individual or community harm, or they may be called at courts of law as experts in public health or specialists in other medical specialties to help judges and juries determine the best solution for a public health problem under litigation.

Language of Medicine and Law

The language of medicine, epidemiology, and law are sometimes very similar. Sometimes, however, more important differences must be noted and interpreted by all and used in common and reciprocal understanding in tort litigation. Let us start with the term *tort litigation*. Do all health professionals called to courts understand them? Are all lawyers aware of some differences that might be crucial in analyses, interpretations, and verdicts? In this light, what are tort, litigation, and liability?

Tort is defined in law as "a wrongful act, other than a breach of contract, that results in injury to another person, property, reputation, or some other legally protected right of interest, and for which the injured party is entitled to a remedy at law, usually in the form of damages."[36] As we can see, the term "wrongful act" here corroborates in general with our medical definition of error and harm as discussed in this book. Usually, medical error as a wrongful act and its consequences is an unintentional error leading to an *unintentional tort*. Not all medical errors lead to tort and not all give rise to liability and a right to compensation.

An *intentional tort* is "a tort committed by one who intends by his (or her) action to bring about a wrongful result or knows that the result is substantially certain to occur."[36] The intentionality of a medical error leading to harm must then be the subject of additional analysis and proof.

Litigation is a legal term meaning in broadest terms "a case or a set of cases discussed collectively"[36]; in the context of this chapter, we can add "at courts of law." More specifically, litigation means (in our context) lawsuits or legal proceedings. Hence, cases of medical error discussed by health professionals in their healthcare and research environment only are not and should not be seen as litigations!

Liability (in our context) means legal responsibility for a tort or breach of contract, that is, in the sense of civil liability rather than of criminal liability.[36] Medical error may produce (lead to) some kind of harm and tort, subject to litigation. Liable individuals and potentially liable components of a wider health system are searched for and accounted for.

Table 8.1[37–61] reminds us of the semantic heterogeneity that may exist among the domains of law, lathology, and medicine, epidemiology, and health sciences. Even basic terms like error, evidence, mistakes, lapses, or causes may have, to a variable degree, different meanings in these areas. This table presents an array of meanings for comparisons; definitions of most terms cannot be in all cases prescriptive. Also, in many cases, important terms are not defined in all three mentioned domains (law, lathology, medicine).

For the moment, in all medico-legal discussions of health topics, a common understanding of terms and their definitions, meanings, and uses by health and law professionals must be reached first case by case and litigation by litigation. As a caricatured example, and just for laughs, we can discuss the term *abduction* with women and men of law. For a physician-critical thinker, abduction means a form of reasoning in which one reasons from observed phenomena to a hypothesis that would explain them. The resulting hypothesis is a possible explanation, and further observation or experiment and reasoning are required to determine whether the hypothesis is justified. (This is like going from data and information taken from a clinical examination and history to a working diagnosis, justified by a more complete clinical work-up in a given patient.) In law, abduction means kidnapping or the tort of luring away, carrying off, or concealing another's spouse, child, or another relative.[36] Culprits are not limited to family members; others might be accepted too. This is only to show how vastly different meanings can be. Fortunately, in most cases, the differences are subtler.

Before tackling error and harm, law and health professionals must always agree first on the meaning of the terms they will use.

General Philosophy and Strategies of Medicine and Law

In essence, physicians are mainly probabilists, and lawyers are more determinists in their final decisions and verdicts.[47] Physicians are accustomed to working and making decisions under varying degrees of uncertainty. Lawyers and judges, despite uncertainties in the legal process must, in the end, pronounce a clear-cut verdict of culpability (liability) or nonculpability (non-liability).

In the spirit of such distinctions, Table 8.2 presents examples of differences in general philosophy and strategies derived from psychology,[62-64] adapted and expanded to medicine and law.[65]

Law Process and Its Stages

Physicians and other health professionals have been introduced also elsewhere in related literature to legal matters and proceedings in the health domain and to what is expected of them in the world of law, litigation, and courts.[65–70]

In law, a lawsuit is a civil action brought before a court in which a party (*plaintiff*), claiming to have been damaged (injured) by a defendant's actions, seeks a legal or equitable remedy. The *defendant* is required to respond to the complaint of the plaintiff. The conduct of a lawsuit is called *litigation*.[71,72] The whole process may be divided into three parts:[67,68] (1) what happens before the trial; (2) what happens at the trial including judgment; and (3) what happens after the judgment (verdict) is pronounced. This may be summarized in an abridged form from several sources[62,65,67-69,72,73] as explained in the following sections.

Happenings and Events before the Trial

Several events (which may vary from one jurisdiction to another) happen before the trial:

■ The aggrieved party (the plaintiff) consults a lawyer, who generally needs to obtain expert medical opinion about the standard of care, whether it was breached, and possibly also about causation to advise whether compensable harm or loss has occurred and if a *valid cause for action* exists.

Table 8.1 Diversity of Terms and Their Meanings in Law, Lathology, and Medicine, Epidemiology and Health Sciences

Law	Lathology	Medicine, Epidemiology, Health Sciences
1. Error		
Error: Syn. to mistake.[37] ... A mistaken judgment or incorrect belief as to the existence or effect of matters of fact, or a false or mistaken conception of an application of the law.[38] ... A mistake of law, or false or irregular application of it, such as vitiates the proceedings and warrants the reversal of the judgment.[38] ... A psychological state that does not conform to objective reality; a belief that what is false is true or that what is true is false.[39] ... A mistake of law or of fact in a court's judgment, opinion, or order.[39] ... Incorrect information including omission of information.[40]	All those occasions in which a planned sequence of mental and physical activities fails to achieve its intended outcome, and when these failures cannot be attributed to the intervention of some chance agency.[31]	The failure of a planned action to be completed as intended or the use of an incorrect plan to achieve an aim.[41] ... Errors may be errors of commission or omission, and usually reflect deficiencies in the systems of care.[48]
Human error	Failures committed by an individual or a team of individuals. (See the Glossary at the end of this book for more formal definitions.)	Operator's error in reasoning, understanding, and decision making about the solution of the health problem and/or in the ensuing sensory and physical execution of a task in medical or community care.
Err: To make an error; to be incorrect or mistaken.[39]		

Failure: being unsuccessful, inadequate, deficient.[36]		Not reaching objectives in any step of medical care such as not obtaining a right diagnosis, not curing the patient, and not improving patient prognosis due to system or human reasoning and deficiencies in decision making, execution, and evaluation.
Mistake: Some unintentional act, omission, or error arising from ignorance, surprise, imposition, or misplaced confidence. ... It may arise either from unconsciousness, ignorance, forgetfulness, imposition, or misplaced confidence.[38] ... An error, misconception, or misunderstanding; an erroneous belief.[39] ... Misunderstanding about the existence of something which arises either from a false belief or ignorance.[49] ... Mere forgetfulness is not a mistake against which the court will grant relief. ... A mistake may be common, mutual, or unilateral, mistake of law and mistake of fact.[55]	Deficiency or failure in the judgmental or inferential processes involved in the selection of the objective or in the specification of the means to achieve it, irrespective of whether the actions directed by this decision-scheme run according to plan.[31]	Failure in clinical or medical research reasoning and decision making.
Lapse: To slip: to deviate from the proper path.[38]	Deficiency or failure in the execution and storage stage of an action sequence, regardless of whether the plan that guided them was adequate to achieve its objective.[31]	Failure in execution and performance of tasks in clinical or community care.

Continued

Table 8.1 Diversity of Terms and Their Meanings in Law, Lathology, and Medicine, Epidemiology and Health Sciences (*Continued*)

Law	Lathology	Medicine, Epidemiology, Health Sciences
2. Harm		
Risk: Possible harm or loss or chance of danger.[37]	The result of error frequency combined with or multiplied by the severity of consequences, resulting from its occurrence.[42]	The probability that an event will occur, for example, that an individual will become ill or die within a stated period of time or by a certain age.[43]
Harm: Injury, loss, or detriment.[39]		Harm includes disease, injury, suffering, disability, and death.[44]
Adverse (adverse event): Anything contrary or that goes against one party.[37]		Untoward incident, therapeutic misadventure, iatrogenic injury, or other adverse occurrence directly associated with care or services provided within the jurisdiction of a medical center, outpatient clinic or other facility.[45] ... Any of the following three: (1) an unexpected and undesired incident directly associated with the care or services provided to the patient; (2) an incident that occurs during the process of providing healthcare and results in patient injury or death; (3) an adverse outcome for a patient, including an injury or complication.[46] ... An injury related to medical management in contrast to complications of disease. ... May be preventable or nonpreventable.[48]

Event			Any deviation from usual medical care that causes an injury to the patient or poses a risk of harm. Includes errors, preventable adverse events, and *hazards*, that is, any threats to safety like unsafe practices, conduct, or equipment.[48]
Accident: An unintended and unforeseen injurious occurrence; something that does not occur in the usual course of events or that could not be reasonably anticipated. … An unforeseen and injurious occurrence not attributable to mistake, neglect, or misconduct.[39]			An unanticipated event—commonly leading to an injury or other harm; in traffic, the workplace, or a domestic or recreational setting.[43]
Incident: Something appertaining or attaching to something else; a feature, characteristic, or concomitant. … Something that occurs in connection with something else, but not as a necessary component or accompaniment. … An occurrence or event. …Pertaining; occurring in connection with.[36] …	Events, processes, practices, or outcomes that are noteworthy by virtue of the hazards they create for, or the harm they cause, subjects.[47]		Events, processes, practices, or outcomes that are noteworthy by virtue of hazards they create for, or harms they cause, patients.[46] … Any deviation from usual medical care that causes an injury to the patient or poses a risk of harm. Includes errors, preventable adverse events and hazards.[48] … An event or circumstance which could have resulted, or did result, in unintended or unnecessary harm to a person or complaint, loss, or damage.[44]

Continued

Table 8.1 Diversity of Terms and Their Meanings in Law, Lathology, and Medicine, Epidemiology and Health Sciences (Continued)

Law	Lathology	Medicine, Epidemiology, Health Sciences
3. Evidence		
Evidence: Written or spoken statement of fact that helps to prove something at a trial.[37] Any species of proof, or probative matter, legally presented at the trial of an issue, by the act of the parties and through the medium of, for example, witnesses, records, documents, exhibits, and concrete objects, for the purpose of inducing belief in the minds of the court or jury as to their contention.[38] ... Something (including testimony, documents and tangible objects) that tends to prove or disprove the existence of the alleged fact. ... The body of law regulating the burden of proof, admissibility, relevance, and the weight and sufficiency of what should be admitted into the record of a legal proceeding.[39] ...The means, exclusive of mere argument, that tends to prove or disprove any matter of fact the truth of which is submitted to judicial investigation.[50] ... That which tends to prove the existence of some fact.[42]		Any data or information, whether solid or weak, obtained through experience, observational research, or experimental work (trials), relevant to some degree (more is better) either to the understanding of the problem (case) or to the clinical decisions made about the case.[1,52]

Best (primary) evidence: Evidence of the highest quality available, as measured by the nature of the case rather than the thing being offered as evidence.[39]	Most valid, among all available options, clinically relevant findings from research, together with medical care expertise corroborating facts.
Direct (positive) evidence: Evidence that is based on personal knowledge or observation and that, if true, proves the fact without inference or presumption.[39]	
Circumstantial (indirect, oblique) evidence: Evidence based on inference and not on personal knowledge or observation.[39] … Secondary facts by which a principal fact may be rationally inferred. … The circumstances must be consistent with the conclusion that the act was committed by the accused and inconsistent with any other rational conclusion.[55]	
Critical evidence: Evidence strong enough that its presence could tilt a juror's mind.[39]	
Medical evidence: Evidence furnished by a doctor, nurse, or other qualified medical person testifying in a professional capacity as an expert, or by standard treatise on medicine and surgery.[39]	See above (our definition of evidence in medicine)

Continued

Table 8.1   Diversity of Terms and Their Meanings in Law, Lathology, and Medicine, Epidemiology and Health Sciences (*Continued*)

Law	Lathology	Medicine, Epidemiology, Health Sciences
Scientific evidence: Testimony or opinion evidence that draws on technical or specialized knowledge and relies on scientific method for evidentiary value.[39]		See above (our definition of evidence for medicine)
4. Critical Thinking, Argumentation, Logic		
Knowledge: What is known to be true.[37]	Justified true belief	Justified true belief. … True belief that has been acquired by a generally reliable process.[53]
Objective knowledge: Facts or information that is generally known by persons in the same situation as the person in question.[55]		
Argument: Discussing something without agreeing; (speech giving) reasons for something.[37] … An effort to establish belief by a course of reasoning.[38] A method of establishing belief using a course of reasoning.[40]	Production of considerations designed to support a conclusion.[54]	A structured and organized dialogue between stakeholders in health and disease with predetermined aim to improve understanding of health problems and to make correct decisions about them.[52]

Syllogism: A form of reasoning in which one draws a conclusion that does not contain the common element of premises.[49]	The inference of one proposition from two premises.[54]	A set of medical or other statements, some of which, the premises are reasons for another statement, the conclusion,[53] like diagnosis or treatment decision.
Premises: That which is put before; that which precedes; the foregoing statements.[38] (N.B. Like premises that precede and lead to a conclusion of a categorical syllogism in logic.) ... "In consideration of premises" means in consideration of matters hereinbefore stated.[38]	One of the propositions from which together the conclusion is derived.[54]	A starting point of medical reasoning and argument, that from which the conclusion is drawn, possibly in combination with other premises.[53]
Inference: A deduction from the evidence presented that, if reasonable, may have the validity of legal proof.[55]	The process of moving from (possibly provisional) acceptance of some propositions, to acceptance of others.[54]	
Claim: An assertion that one is entitled to something. An assertion of facts that, if true, would legally entitle the claimant to judgment in a civil case. An apparent or actual right to receive something by way of a lawsuit.[35]		As in logic and critical thinking: One of the six elements in the contemporary layout of arguments. The proposition at which we arrive as a result of our reasoning, or which we defend in argument by citing our supporting grounds.[53]

Continued

Table 8.1 Diversity of Terms and Their Meanings in Law, Lathology, and Medicine, Epidemiology and Health Sciences (Continued)

Law	Lathology	Medicine, Epidemiology, Health Sciences
5. Causation		
Cause (causation): Something that produces an effect or result. ... The plaintiff must prove causation.[39] ... That which produces an effect and includes any action, suit or other original proceeding between a plaintiff and a defendant and any criminal proceeding by the Crown. A suit or action[49] ... That which affects a result; a motive or reason. In law, cause is not a constant and agreed upon term.[54] ... The relationship between an act and the consequences it produces. It is one of the elements that must be proved before an accused can be convicted. [51] (N.B. Language of being "convicted," however, is not used in civil proceedings.[56])	An antecedent set of actions, circumstances, or conditions that produce an event, effect, or phenomenon. A cause may be proximate (immediately precede) or remote (a factor predisposing to) the event, effect, or phenomenon.[44]	As in *Encyclopaedia Britannica:* That factor which is possible or convenient for us to alter to produce or prevent an effect ... This concept contains two components, production of an effect and understanding of its mechanisms.[1] ...In general: An event without some subsequent event would not have occurred or because of which it occurred. ... An agent or act that produces some phenomenon (effect)[53]
Proximate cause ("not too remote" cause[56]): A cause that is legally sufficient to result in liability. A cause that directly produces an event and without which the event would not have occurred.[39] ... That which is a natural and continuous sequence, unbroken by any new independent cause, produces an event, or without which the injury would not have occurred.[55]		

Direct cause: The active, efficient cause that sets in motion a train of events that brings about a result without the intervention of any other independent source. Often used interchangeably with proximate cause.[55]		Superseding cause: An intervening cause that is so substantially responsible for the ultimate injury that it acts to cut off the liability of preceding actors regardless of whether their prior negligence was or was not a substantial factor in bringing about the injury complained of. ... Novus actus interveniens.[55]
Remote cause: A cause that does not necessarily or immediately produce an event or injury.[39] (N.B. Just because a cause is remote does not mean that law will not attach liability; there need not be an immediate temporal relationship.[56r])		Close to the meaning of the latent source of error and harm, a remote event or action, product of a system, mostly machine, function, interaction/ organization of technical, environmental or social factors. N.B. The remoteness or proximity to action of error and harm source remains unspecified for the moment. ... Latent error (latent failure) is a defect in the design, organization, training or maintenance in a system that leads to operator errors and whose effects are typically delayed.[48]
Superseding cause: An intervening cause that is so substantially responsible for the ultimate injury that it acts to cut off the liability of preceding actors regardless of whether their prior negligence was or was not a substantial factor in bringing about the injury complained of. ... Novus actus interveniens.[55]		

Continued

Table 8.1 Diversity of Terms and Their Meanings in Law, Lathology, and Medicine, Epidemiology and Health Sciences (*Continued*)

Law	Lathology	Medicine, Epidemiology, Health Sciences
Intervening or supervening cause: A cause that comes into operation in producing the result and is a later event after negligence of the defendant.[55]		
6. Domain of Law		
Malpractice: Improper or unskilful conduct on the part of a medical practitioner that result in injury to the patient. ... Generally describes professional misconduct or negligence on the part of a person delivering professional services ...[55]		Any professional misconduct, unreasonable lack of skill of fidelity in professional duties, or illegal or immoral conduct. ... It is one form of negligence (see below). ... a bad, wrong, or injudicious treatment of patient professionally; it results in injury, unnecessary suffering, or death to the patient. [27]
Negligence: Omitting to do something that a reasonable man would do or the doing something which a reasonable man would not do.[55] ... In tort law: A conduct involving an unreasonable risk of injury or loss to others; conduct that fails short of the degree of care that a reasonable person would have exercised in the same circumstances.[35]		An incident causing harm, damage or loss as the result of doing something wrong or failing to provide a reasonable level of care in a circumstance in which one has a duty of care.[44]

Negligence (contributory): Conduct on the part of the plaintiff that falls below the standard of care to which he or she should conform for his or her own protection and that, when combined with the defendant's negligence, was a legally contributing cause bringing about the plaintiff's harm or injury.[55]	
Tort: A civil wrong for which a remedy may be obtained, usu, in the form of damages; a breach of a duty that the law imposes on everyone in the same relation to another as involved in a given transaction. ... The branch of law dealing with such wrongs.[39]	
Litigation: A controversy in a court; a judicial contest through which legal rights are sought to be determined and enforced.[55]	
Plaintiff: The one who initially brings the suit; a person who brings an action; a defendant who brings a counterclaim.[55]	In health-related cases, usually patients or civic groups (parties) and their legal representatives soliciting the solution of a health problem. (N.B. More specifically, through a civil claim for damages compensation for injuries that the plaintiff claims were caused by the defendant's breach of standard care of a health problem.[56])

Continued

Table 8.1 Diversity of Terms and Their Meanings in Law, Lathology, and Medicine, Epidemiology and Health Sciences (Continued)

Law	Lathology	Medicine, Epidemiology, Health Sciences
Defendant: The party responding to the claim of the plaintiff.[55]		In health-related cases, usually a physician or other health professional, health organization or structure called to respond to a plaintiff's complaints.
Case: An action, cause, suit, or controversy at law or in equity; a trial; the evidence and argument on behalf of the parties.[55]		Case (clinical): Form of a professional or scientific communication to describe, analyze, and explain a clinical event (patient or clinical or community care situation).[57]
Burden of proof (onus probandi): The duty or onus to prove one's case.[55]		
Legal burden or burden of proof simpliciter: The obligation of a party to meet the requirement that a fact in issue be proved (or disproved) either by preponderance of the evidence or beyond reasonable doubt as the case may be. In civil law, a plaintiff must prove his or her case on a balance of probabilities, whereas, in criminal law, the Crown must prove each and every element of the case beyond a reasonable doubt.[55]		

| *(Professional) Standard of care:* The level or nature of conduct necessary to avoid liability for negligence, malpractice or breach of fiduciary duty.[36] ... The standard by which conduct would be measured ... in terms of what can reasonably be expected of people.[58] ... In tort law, the standard of care is the degree of prudence and caution required of an individual who is under the duty of care. A breach of the standard is necessary for a successful action in negligence. The requirements of the standard are closely dependent on circumstances.[59] | *(Medical) Standard of care:* The level at which the average, prudent provider in a given community would practice. It is how similarly qualified practitioners sould have managed the patient's care under the same or similar circumstances. The medical plaintiff must establish the appropriate standard of care and demonstrate that a standard of care has been breached.[60] ... It specifies appropriate treatment based on scientific evidence and collaboration between medical and/or psychological professionals involvement in the treatment of a given condition.[59] ... The benchmark against a doctor's actual work. ... A formal diagnostic and treatment process a doctor will follow for a patient with a certain set of symptoms or a specific illness. That standard will follow guidelines and protocols that experts would agree with as most appropriate, also called "best practice."[61] |

Table 8.2 Differences between the Philosophy and Strategies of Medicine and Law

Medicine	*Law*
Training in a clinical science environment.	Training in an environment of problem solving while balancing knowledge, facts, and evidence brought in by all parties involved, in the spirit, culture, values, and laws of the nation and society.
Depends on creativity.	Depends on precedents. (N.B. but also on creativity (law develops), statutes, and regulations.[56])
Embraces, accepts, and encourages scientific method and its results.	Authoritative declarations are accepted (but the burden of proof in civil proceedings takes into account the balance of probabilities[56]).
Collects, analyzes, and produces objective (preferably bias-free) data and information as a basis for acceptable evidence.	An adversarial method is used to decide what "facts" are accepted.
Nonjudgmental descriptions of what is and was seen are reported.	Moral values prescriptions for how to behave are used.
Medical decisions are based on and influenced most often by the physician's understanding of large concurrent sets of cases (and controls), cross sectionally or longitudinally assembled and studied.	The focus is on a historical line of cases appearing at courts of law. Decisions are made with a variable and acceptable degree and margin of error. Problems, as cases to solve in research, require most often searching for them as well as for individuals as clinical cases.

Source: Reworked with modifications from Haney, C., *Law Human Behav.,* 4, 3, 1980; Dalby, J. T., *Can. Fam. Physician (CFP),* 53, 2007.

- The demand for the case to be heard is formulated as a "lawyer's letter." A *statement of claim* (some jurisdictions will begin actions using writs) is *filed* with the clerk of the court.
- A sort of *prelitigation settlement discussion* may be held between the plaintiff and the defendant and their lawyers to see if some kind of settlement can be reached without going to trial.
- Both parties can engage in a series of procedures, a process of "*discovery.*" During such processes, both parties may seek further details about the opposing party's case to avoid "surprises" at the court.
- The ensuing *Motion for Summary Judgment* is filed to assert if there is a need for a trial or if a legal decision can be made at this moment as a kind of "advanced ruling."

Physicians as defendants or expert witnesses may participate in this kind of pretrial activity to help lawyers and the court further speed up the decision-making process and its results about the case.

At the Trial

This is generally the moment in the trial process most familiar to health professionals, including novices:

- The *jury* (not used nearly as frequently in Canada as in the United States) is selected, and the *date* for the trial is set.
- *Opening statements from both parties* are presentations to the jury about what should be expected.
- The plaintiff (his or her lawyer) presents the *plaintiff's case* first, which is then examined and cross-examined by parties (i.e., the plaintiff and the defendant).
- The *defendant's case presentation* follows, equally subject to cross-examination.
- A *motion for directed verdict* may be filed, quite similar to a *Motion for Summary Judgment* (see the pretrial period) in its legal effect, but it can be made only at the end of the trial.
- *Closing arguments* are presented by lawyers for both parties (i.e., the plaintiff and the defendant).
- Both parties are permitted to propose *jury instructions* (i.e., legal rules to be applied in this case).
- The *jury's decision* is made and reported as its verdict to the trial judge.
- The trial *judge pronounces his or her verdict.*

After the Trial

After the verdict is pronounced at the end of the trial:

- *"Post-trial motions"* may be filed, mainly by the defendant's party to convince the judge that something else is appropriate in the verdict or in its parts.
- If any party is unhappy with the verdict, the case may be sent, depending on the law and rules of court in the jurisdiction about appeals, to an appellate court for further review. This process is called an *appeal*.
- An *enforcement* of the final judgment is implemented barring the plaintiff from bringing the same or similar claims against the defendant or from relitigating any of the issues, even under different legal claims or theories[71] and ensuring that the defendant will comply. (Relitigating of any of the issues is not "enforced" in any active way unless there is a subsequent legal proceeding.)

Physicians may be consulted and called to testify as expert witnesses by the party that has requested their service (either the plaintiff or the defendant) at any stage of the litigation process or by the court (judge) at the moment of the trial. They may be required by their inviting parties to present their views and be counterinterrogated by the opposing party. The judge may require answers to any questions that are within the expertise of the health professional as an expert witness to facilitate him or her in making the best possible verdict on the basis of the evidence presented. The expert witness must be as objective as possible and not be a partisan advocate. In the case of medical error, questions may arise about the nature of the error itself, its causes, and consequences if the error leads to harm. Cause–effect associations are most often in focus. Webs of causes, webs of consequences, or rebuttals are frequently scrutinized, and the expert witness must expect them.

Physicians as defendants in malpractice litigations may expect the same, although this is not the only focus. For defendants, much of the attention may be on "what happened."[56]

Cause–Effect Relationships in Medicine and Law

Linking epidemiological and legal reasoning developed during the past half-century.[18,30,74,75] Causation and causal inference in the health domain was adapted and expanded into its current form and content[1,76] from basic philosophical considerations,[53] particularly those of John Stuart Mill and used also by Austin Bradford Hill[74,75] and Jenicek's *Foundations for Evidence-Based Medicine* (which provides an expanded list of references)[1] for health sciences[76,77] and even paralleled with legal reasoning in criminal law.[78]

However, the medical and legal emphasis and focus on various aspects and criteria of causality may differ. Sagall[79] stresses among other things that medicine pays particular attention to the etiology of disease, while law is often interested in the more proximate cause of injury, disability, or death. While physicians identify causes of an entire disorder in medicine, lawyers may concentrate particularly on participating, hastening, or aggravating causes of some pathology. Whereas a scientific proof of causation is required in medicine, the law considers probability and not absolute certainty for causation. Ultimate answers can be deferred in medicine, pending new scientific advances, but in law, the issue must be decided when presented.

Mayhem or murder in criminal law and morbidity, mortality, and causality in medicine apply many semantically close terms and criteria like *premeditation* (causal vents precede the disease like error precedes harm or its causes precede error), *motivation* (the role of error or its causes must make biologic, contextual, common and other sense), *proof of guilt* (the proof of causation must be established beyond reasonable doubt or role of chance), and others.[78,1] In negligence law, the defendant's *motive* is not relevant to legal liability; the issue is the defendant's *conduct* (i.e., acts or omission).[56]

In various national legislations, error is not always synonymous with criminal offence or civil liability, both being separate entities in the meaning of law. In Italy, error is not enough to substantiate a hypothesis of a criminal offence (civil liability); not even the occurrence of a harmful effect is sufficient. Olivetti et al.[4] point out that an indispensable feature is the causal criteria relationship between the error (or misconduct) and the harm. Harm must be directly caused by the medical act with an uninterrupted causal link and not from contingencies rooted in the disease itself or intervening extraneous causes.

In civil law (negligence), the error does not have to be the sole cause of the injury, just a cause.[56] In civil lawsuits pertaining to cause–effect relationships, the burden of proof depends on the balance of probabilities. In criminal law, proof must be beyond a reasonable doubt, that is, "with a high degree of probability near to certainty" that the harmful event "…would not have occurred or would have occurred at a significantly later time or with less damaging effects…" had the physician acted differently.[4] Both contractual liability and extracontractual liability are of interest to courts. They are evaluated at the human and system level as well, and physicians and the health system may be liable.

Chapter 5 also partly covers the challenge of causality in the context of medical error.

Physicians' Roles in the Judicial Search for Causes

As already mentioned, physicians are present at courts for two major reasons:

1. They are called to courts as *expert witnesses* to bring a judgment and explanations to events in the domain of human biology, health and disease, and medical care.
2. They are brought to courts by various plaintiffs (e.g., patients as individuals or groups of patients, and other concerned individuals or groups in the community and their lawyers) in individual or class actions becoming *defendants* (alone or represented by their lawyers); they are accused by plaintiffs of wrongdoing (i.e., errors leading to harm) and are asked for compensation for short-term and long-term effects alleged to result from their professional conduct and doings possibly being erroneous and leading to a detrimental impact on patient or community physical, mental, material, and social well-being.

In court, physicians are often asked to answer at least five kinds of questions:

■ Is there a cause–effect relationship in general between a presumed medical or nonmedical factor (e.g., treatment, pollution) and a particular disease (cancer, handicap after surgery, becoming emotionally, socially, and professionally dysfunctional after psychiatric care)?
■ Is this specific case of the patient's condition a result of medical error?
■ Is such a medical error imputable to a physician, other health professional, healthcare group, or health institution?
■ Are there (within the physician expert witness' area of expertise) other civic bodies, industries, work environments, or lawmakers responsible or coresponsible for such errors and their consequences?
■ Did some particular factor cause a health problem in this particular individual or group of individuals in a class action?

Similarly, we may raise questions like the following:

■ Is there a cause–effect relationship among some factor, characteristic, or event leading to error (or harm)? Statistically speaking, error in this question is a consequence—a dependent variable under study.
■ Does this particular error cause harm and other undesirable outcomes, or has it done so? Statistically speaking, error in this question is a presumed cause or independent variable under study.
■ In law, what was the standard of care, and was it breached?[56]

Levin[80] defined a general approach to these proofs as follows: "...The standard of proof in the courtroom is similar to that in clinical medicine. To be indicted as a causative factor in an illness, an agent must be shown to a reasonable degree of medical certainty to cause or substantially contribute to that illness. This standard of proof is true for infectious and toxin associated diseases in clinical medicine...."

More specifically, two levels of decision must be made. In civil lawsuits, it must be demonstrated that a particular agent or act is a *more probable than improbable cause* of a given health problem (i.e., "more probable than not" or "this one rather than the other one"). In criminal law, the standard is proof beyond a reasonable doubt is enough and must be demonstrated beyond a reasonable doubt that a health problem being litigated or that gave rise to the litigation (paralysis, coma, death) is *due to* a given factor (cause). In other terms, "it cannot be the other one."

Hoffman[81] summarizes this point as follows:

> Whereas scientific proof and statistical significance rest by convention on demonstration of a 95 per cent or greater probability that the results of an experiment or investigation were not due to chance, in civil tort cases the most widely used standards of proof require a "preponderance of evidence," "more likely than not" a "50.1 per cent or more probability," "but for," or "reasonable medical certainty." Probable is defined as "having more evidence for than against." "Preponderance of evidence" is defined as "evidence of greater weight or more convincing than the evidence which is offered in opposition to it.... "Preponderance" denotes a superiority of weight." In criminal cases, the usual standard of proof is "beyond a reasonable doubt." Therefore, quantitative epidemiologic data must be expressed more qualitatively in the courtroom...."

In general, it must be drawn from up-to-date knowledge of specific evidence and research. The question of whether the criteria of causality (as reviewed in this chapter) were met[82,83] is also important. For Evans,[83] criteria of causality are not fixed, but they evolve with technology.

Medicine operates not only with *hard* data like blood counts or results of cultures but also with *soft* data, which is harder to define in operational terms, to measure, and to quantify, for example, with observations in psychiatry or in the area of pain. First and foremost, were all variables (e.g., cause, effect) well defined qualitatively and quantitatively? Was exposure well defined and quantified? Are health problems defined satisfactorily from the point of view of clinimetrics? What is fatigue? What is malaise? What is a lack of concentration?

Frequently, very soft and nonspecific clinical data are presented as evidence, and only physicians can help the court make the "right" decision.

Is the Causal Link under Review Strong and Specific Enough?

In civil law, if one cause is known, its etiological fraction should be superior to 50%. Hence, if more etiological factors are known, should the etiological fraction of the factor under scrutiny be superior to any other known?

In criminal law, an absolutely specific relationship is sought in terms of an attributable fraction approaching 100%. For example, based on Cole and Goldman's review,[84] in the study of occupational exposure to vinyl chloride monomers as a cause of liver cancer, a 200.0 odds ratio (giving an approximately 99.5% attributable fraction) was found. A 2.0 odds ratio relating benzene occupational exposure to bone marrow neoplasia giving a 50% attributable risk is less convincing than the former.

Similar attention must be paid to *all other criteria of causality* rather than to the strength of a cause–effect relationship in terms of relative risks of odds ratios. Those criteria are reproduced again in Chapter 5 and elsewhere.[1] In law, some factor as a possible cause of harm needs only to be a cause; it needs not be the only cause of harm.[56]

What Is Sufficient and Best Proof for Physicians and Lawyers?

Physicians and lawyers differ in these matters.[85,86] Let us see how they differ specifically in reasoning about what is proof and what isn't.

What Do Physicians Think?

In evidence-based medicine, the hierarchy of evidence focuses so far mainly on a cause–effect relationship at the disease level or at the level of disease in a particular group of patients. The clinician who uses the best evidence then evaluates if such best evidence applies to a particular patient and setting of clinical care. Ideally, but not always, the build-up of evidence progresses from a basic laboratory inquiry → to animal research that produces → other ideas, editorials, opinions → reflected in case reports and case series → that serve as a basis for case control studies → to generate ideas for observational cohort studies → for randomized controlled, multiple-blind clinical and community trials → and systematic reviews and meta-analyses of their results, those latter being considered sometimes as the best evidence (for the problem as a whole). Applications of findings in groups to individual patients are challenging. For example, more strictly speaking, results of a clinical trial would be most suitable only to a patient who would be eligible (due to, e.g., his or her characteristics, disease state) to be enrolled and participate in such a trial and whose characteristics are comparable to those individuals in the trial who responded well to the treatment, a consideration still not used often in practice.[1]

Clinical case reports or case series reports in the absence of any comparisons represent in evidence-based medicine a low level of cause–effect proof such as the effectiveness of a particular treatment. However, if any other levels of proof cannot be reached for whatever reason, the following lowest level of proof is the best proof "given the circumstances."

Also, in the case of rare events, adverse effects in clinical pharmacology, or cases that cannot be reproduced for ethical and other reasons or if suitable control groups cannot be found, single clinical cases and case series are what is left, and they must be analyzed and interpreted to their maximum.

What Do Lawyers Think?

First and foremost, lawyers are interested in causes (or at least in one cause, such as that of a defendant) of problems in a particular individual, given circumstances and specific settings and how such a specific case is supported by the broader knowledge of causality of the health problem in its entirety and as a whole in the population and given environment. Consequently, for example, if nothing else is left or possible, individual clinical case reports and case series may be seen as a more important demonstration of evidence than in the general hierarchy of evidence across the whole health problem and its management in research and clinical care. Narratives, case studies, and case reports are much more valued evidence, especially in the absence of other, stronger evidence because they help the lawyer or judge make the best decision in an individual case and for a specific individual.

Some tort decisions in courts based on causation supported by relative risks or odds ratios of 2.0 and more may still be wrong because relative risks or odds ratios reflect the strength of a causal association only, while attributable risks or etiological fractions reflect its specificity and not other criteria of causality. They may or not follow in the same direction. Technical note: We have already dealt in the past with levels of relative risk worthy of attention. This problem also depends on confidence intervals of relative risks saddling or not the value of 1.0.[1]

Thus far, legal grading in the hierarchy of evidence is different from evidence grading in evidence-based medicine. It may be based, with an increasing level of evidence for law, on proof standards such as in the following example:[85,86]

Regulatory Standard	*Level of Evidence:* –
Legal-civil evidence based on a *"more likely than not"* standard and consideration.	* (lowest)
Legal-civil evidence based on *"clear and convincing"* evidence.	**
Legal-criminal evidence where proof *"beyond a reasonable doubt"* is necessary.	***
Scientific evidence based on some degree of *"irrefutable"* evidence.	**** (highest)

Such a hierarchy of evidence still must prove its worth in wider practical use. Recent Supreme Court of Canada jurisprudence confirms that there is only one standard in civil cases (i.e., proof based on the balance of probabilities). Some hierarchy of evidence may still prevail in U.S. state jurisdictions and elsewhere.[56]

The same medical and legal authors[85,86] state the following:

> The most essential evidence in medicine is the patient story. (N.B. In what sense?) In law, eyewitness testimony (i.e. a case report) can meet the highest legal standard of proof, beyond a reasonable doubt (as required in criminal cases, not in civil cases). Medical evidence does not often meet the scientific standard of proof; and, as in law, it should be judged by a standard of proof appropriate for the fact or point in question. An anecdotal case report can provide evidence of probative value, just like eyewitness testimony in a murder trial. And it can be similarly tested by second opinions, re-examination, laboratory tests, and follow-up.[85]

The value of single cases as causal proof is also questioned by clinical pharmacologists and homeopaths who often work with rare cases and events. In clinical pharmacology, anecdotal or single clinical case reports (e.g., adverse effects, errors, harm) may bring good evidence in specific cases and may be paralleled with reasoning in the law domain if the following[87] is found:

- Extracellular deposition of drug or metabolite (culprit caught at scene of crime).
- Intracellular deposition of drug or metabolite (culprit caught at scene of crime).

- Specific location or pattern of injury (culprit seen committing the crime, despite fallibilities in eyewitness identification).
- Physiochemical dysfunction or tissue damage (culprit incriminated by recreating the crime scene).
- Infection related (culprit's DNA found at scene of crime).

In the domain of alternative medicines, Kiene[88,89] proposes within his "cognitive medicine" scope additional (and often similar to clinical pharmacology) criteria of causality based on the observation of infrequent cases of unique complex responses, and various time, space, morphology, intensity, and other "patterns," "correspondences," or "fits" together with some degree of plausibility, critically reviewed elsewhere.[90] We still do not have enough experience regarding how such considerations might apply to unique cases of error and harm and prove themselves useful along with the formal clinical epidemiological approach. This is worth exploring in some (but not all) cases given their epidemiological similarity with other clinical events appearing in limited numbers and handled with similar logic by health and law professionals.

In law, then, the strength of evidence may differ from a "medical" demonstration of evidence and its uses. Ideally, an agreement of medical and legal parties involved is necessary on a case-by-case basis.

Disease versus Individual-Case Causes: Error as an Entity (in General) and in Specific Cases

Rose[91] rightly distinguishes causes of disease incidence—that is, causes of disease as a mass phenomenon in the community, and causes of a particular case. High national salt intake may be responsible for an important occurrence of hypertension in such a community in general, but what is the reason for developing hypertension in a particular patient? This demonstration is more difficult, causes of cases are less known than causes of incidence, and, presently, conclusions must be based on rigorous judgment and inference from knowledge of causality at the community level.

Ultimately, physicians giving advice to courts must realize that courts are certainly interested in two elements: (1) in a *general truth* (e.g., that a particular system or individual error risks, in general, may lead to (2) some *specific harm in some specific type of individuals) and setting of medical care* in solving the issue before them. However, decisions are made for the specific and precise case of the plaintiff—that is, a particular error occurring in relation to a specific individual or organization's harm.

So, the issue goes from general to specific. Do studies giving a general proof apply to a specific individual? This is similar to the question: "Does this research result apply to my patient?" Did this error occur and cause the plaintiff's harm? General rules of admissibility are partially listed in Chapter 4 with regard to the theoretical eligibility of a particular individual to studies of evidence brought to the attention of the courts.

Litigating the Argumentative Way

Essentials of logic and critical thinking in their general traits and structure are common to medicine[52,53] and law.[21,92] Inductive and deductive reasoning and using categorical syllogisms in argumentation are just some of the ways to ask questions and solve problems in litigations.

Claims discussed at courts and presented to the judge by lawyers, plaintiffs, and defendants are conclusions of arguments, as statements based on other statements, like premises in categorical syllogisms or elements of Toulmin's modern argumentation.[93] (For details see Chapters 2 and 6 or related literature.) All statements and applicable questions may be hidden in natural language, and physicians at courts should be able to understand and evaluate them as well as to detect and correct fallacies.[94] As expert witnesses, physicians are expected to support their views by means of classic and modern argumentation.

For example, a claim is raised such as, "The technical error during the operation led to unexpected harm (long-lasting handicap) in this patient." Once the error and the consecutive harm are clearly defined in operational terms, such a conclusion of an argument must be supported by its grounds, warrant, backing, qualifier, rebuttals, conclusion, or claim:

- *Grounds* are the clinical and other data and information on which the claim is based. Grounds are specific facts relied on to support a given claim or conclusion.[53] A description of "what happened" often contains detailed grounds. For lawyers and judges, grounds may be facts and evidence presented by any party during the litigation process.
- *Warrant* is a more general framework for a given argumentation. It offers a general rule or reference understanding of the problem that allows the inference of the claim from grounds of the type we have adduced.[53] It is a presentation of our understanding of this type of error and harm and our conclusions about it. In medicine, the warrant may be drawn from experience and bibliography search. For lawyers and judges, it may be based on experience drawn from past litigations.
- *Backing* represents the body of experience and evidence to support the warrant.[53] Broad experience with the entire clinical process and care of

this type of individual and pathology, modes, and ways of its medical and surgical management and their expected and other possible outcomes should offer a better and wider understanding of what lead to the argument conclusions (claim). Again, for physicians, backing in research or on clinical rounds and bedside teaching may be based on literature searches. For lawyers and judges, case studies as references to their solution of the present case under litigation play a similar role.

■ *Qualifier* is (ideally) a quantified expression of our certainty about the strength conferred by the warrant on the inference from grounds to claim, and thus the strength of support given to our conclusion by the grounds we offer.[53] We say "how sure we are" that our claim of error and harm in this specific context of the case is correct. Realistic and medical evidence supported qualifiers are perhaps among the most challenging elements in argumentation. Lawyers and judges in their argumentations must often be more unequivocal than probabilistic.

■ *Rebuttals*, defined as extraordinary or exceptional circumstances that would undermine the force of supporting grounds (premises), specify the situations and conditions in which our conclusions (claim) do not work or are not valid.[53] In clinical medicine, contraindications of medical or surgical procedures or drugs carry the meaning and spirit of rebuttals. Here, we focus more on the result of the whole argumentative process. In law, cases coming back to courts after verdicts may be based on rebuttals. Besides the qualifier, an evidence, warrant, and backing-supported rebuttal is one of the most challenging elements in legal argumentations.

■ *Conclusion* or *claim* is a result of the argumentative process containing the five previous elements. It can be built prospectively from the elements in question or can be analyzed retrospectively; that is, if the conclusion is that, how the previous elements support it, what is the quality and completeness of evidence, what and where notable weaknesses and contradictions (rebuttals) are found.

What might be a conclusion or claim of an argument in the medical error domain? Here are some examples:

■ This is an error (and of what nature).
■ This error was caused by malpractice.
■ This community is sick due to the local industry.
■ The misuse of the surgical instruments during the operation led to complications, protracted recovery, and a long-term handicap of our client (the patient as plaintiff).

Claims are endpoints of the modern argument that are subsequently analyzed by the opposing party or lawyers and the judge.

Because medical experts may play the role of plaintiffs, defendants, or both, medical experts are expected to (and will) provide meaningful and truthful claims supported by high-quality arguments. They must be ready to present courts with all such elements to facilitate decision making by the courts. Opposing parties must master the argumentation and the quality assessment of evidence that they contain allowing them to react and offer their views of the claims (conclusions of arguments) presented to the courts. Toulmin's ways of argumentation[55,93] appear for now more familiar in medicine and other health sciences than in the current experience and uses in litigation. The legal field has, however, a longer tradition of using logic and critical thinking.

Disclosure of Medical Errors: Working in Law and Epidemiology with What Is Available

Obtaining an epidemiological portrait (description of occurrence) of error and harm, exhaustive, representative, comparable, and otherwise valid as a reference remains a challenge when dealing with health problems at courts. Creating such a portrait is a kind of "epidemiological disclosure" (i.e., gathering pertinent information). "Disclosure" or "discovery" in the legal world means a set of procedures by which each side in a case may obtain pertinent information from the other.[36] Both desire, by different and often complementary means and for different purposes, the best and most complete picture of the problem.

Understanding what happens in general is vital both for lawyers and physicians. For lawyers, the best possible medical error disclosure brings to their attention not only cases to be dealt with but also a pool of cases to which they may refer in judging specific cases in their broader context. For epidemiologically minded physicians, the degree and completeness of disclosure of medical error is an essence of the epidemiological surveillance of error and harm. Grasping the occurrence of errors, hypothesizing and analyzing their causes and considering error and harm control measures depend on it.

The current problem of disclosure is also magnified by the fact that current legal rules, directions, and mandatory procedures in the health system and the social structure (with the exception of the qualified privilege legislation, perhaps) do not encourage the disclosure of error.

Disclosure of human error, both active (operators at the sharp end of the action) and latent (accumulated with system errors in the past, at the blunt end of the action) and system error as well weigh heavily psychologically, emotionally, and responsibly on the health professional and create an additional social and administrative burden on the health system of clinical and community care. Gelderman's[95] personal experience as a surgeon and emergency physician losing a patient shows that communication, other individuals in the system, and the culture of medical mistake management are all involved in error and harm to the patient.[87] The reality, however, goes beyond individuals.

Barriers to the disclosure were well reviewed by Moskop et al.[96] These authors stressed several important sites of such barriers:

- The ethical responsibility to the patient. Physicians should consider if the harm to the patient was the result of error or if the patient's prognosis might be altered by error and its consequences.
- Within the health and academic system, rules of disclosure are not uniform, and there may be gaps in the follow-up among institutional, home, and community care and communication.
- After the fact, reaching the patient may be difficult.
- Organization of work in the healthcare system may be encumbered by required and missing quality and quantity of care; available technologies; interprofessional communication; and working conditions such as stress, fatigue, and ergonomics, interpersonal relations between healthcare professionals, workload, and material conditions.
- Physicians may consider their errors as a demonstration of their inadequate training, lack of knowledge, right attitudes and skills, erosions of patient trust, or frustration by some lack of long-standing relationship with the patient.
- The recording of the physician's error in his or her personal file may be related to his or her professional performance as a whole.
- Disclosures incite patients to take legal actions against the physician and his or her workplace. Elimination of some current legal barriers should help.
- Fear of publication in both professional and community media, depending also on the extent of the harm to the patient.
- The effect of disclosing errors in liability risk remains unclear.
- Disclosure of errors made by another is a challenge and distinct problem.
- More formal and complete training of physicians in prevention and control of medical error and harm still remains an unreached general objective.

These authors conclude rightly that physicians have a professional responsibility to communicate truthfully with their patients, including errors made. However, "...truthfulness may require that physicians disclose information about previous errors because that information may play an important role in patient decisions about whether to seek care from another physician or hospital or whether to seek redress if harmed."[96]

Qualitative research shows that both patients and their physicians desire full disclosure of a medical error. Focus groups in this research were patients, physicians, residents, nurses, and administrators. Influence on the decision to disclose a medical error includes institutional culture, provider and patient factors, and error factors as well. For harm-producing errors, significant barriers at the level of the provider and institutional culture are hindering disclosure.[97]

Until the problem of disclosure of error is solved, the value and usefulness of epidemiological information and inquiry in error situations remain more than limited. It may be better in some critical cases, but, for the rest, epidemiological inquiry in error means inquiry on what was reported only (whatever its distortion and nonrepresentativity of the problem may be).

A Difficult Mix: Medicine, Ethics, and Law

Medical error and harm are events that are not necessarily synonymous with error and can be contrary to ethics or law. Accommodating this is not always easy. To illustrate such a challenge, let us examine the problem of often life-saving transfusions to patients whose faiths forbid them to receive human blood.

The following 2004 case report of error[98] comes from the U.S. Agency for Healthcare Research and Quality *Case Archive* at the *webM&M Morbidity & Mortality Rounds on the Web*.[99] We have reproduced it, with permission, in its entirety in Appendix 3. In short, a woman severely injured in a collision rejects a blood transfusion for religious reasons. Her relatives persuaded the physician to transfuse her, and she received two units of blood and survived. The Emergency Department staff reversed its decision not to transfuse her according to her wishes when she lost consciousness, and her parents supported the action.

Was this an error in medical decision making? Was the patient harmed? She was not and would not have survived without the transfusion. Physicians did not err in their clinical decision making. On the other hand, due to ethical and legal considerations, this health crisis might be viewed as an ethical and legal error. In the United States, patients have the constitutional right to accept or refuse care and in the eyes of the law, whether the care provided saved the patient's life is irrelevant. The situation may differ elsewhere.

Another case[100] occurred in the Canadian province of Manitoba in 2006. A 14-year-old girl was admitted to the hospital for internal bleeding caused by Crohn's disease. Since she was younger than 16 years old, the cut-off age for medical decision making, the patient's parents given responsibility to make the decision. They objected to the transfusion. Manitoba's Child and Family Services sought and obtained a court order for her to receive a blood transfusion; she was given a life-saving transfusion of three units of blood. Her lawyers argued that her rights under the Charter of Rights were violated, and the matter was brought before the Supreme Court of Canada. In the 2009 verdict, the court upheld the constitutionality of the Manitoba Child and Family Services Act, and its decision said that minors should be given the opportunity to show that they are capable of making their own medical decisions, though they may be overruled by the court if their lives are in danger. Decisions of the Manitoba provincial bodies were not an error in legal terms. A medical error was not made either.

Both examples show that considerations of error or harm may vary. Moreover, legal views and decisions may vary from one jurisdiction to another. In extremis, should a clinician commit a medical-decision error by treating and harming the patient to make his or her actions ethically acceptable and in keeping with local law requirements?

The interface between error and harm in the eyes of a health professional, ethicist, and lawyer still remains an open chapter. For the moment, consensus of all stakeholders in care, including patients, relatives, and possibly others, must be sought on a case-by-case and community-by-community basis.

Conclusions

Physicians are still ill prepared for the challenges of tort litigation regardless of the role they play in the courtroom. Nguyen and Nguyen pointed out various sources of error when dealing with specific problems and subjects of care such as abdominal, chest, leg, flank, or low back pain, injury, and others and what might be the legal issues related to them.[101] They also indicated elements that might generate errors (and lead to litigation later on) and that physicians should therefore pay attention to in their daily practice and clinical care process before going to courts:[102] progress notes and emergency medicine practice (e.g., confidentiality, aggressive treatment, managing patients against their will, rape and abuse management), among others. Whatever the results or outcomes of medical care, they are subject to critical thinking, analysis, and interpretations as outlined in this chapter in the interfacing world of medicine and law.

Remember that in a civil lawsuit the burden of proof rests with the plaintiff.[13] If this were generally known, there would be fewer court cases concerning health problems of individuals or communities. Fortunately, in Canada, there are relatively few medical malpractice lawsuits, let alone trials; their pattern of decline is notable.[56]

Inevitably, expert witnesses are brought to the court both by the plaintiff and the defendant. They are recruited to one side or another no matter their objectivity. The work ethic must be scrupulously respected, as in all fields of epidemiology; its rules are now available in literature.[103]

The study of errors is in even more of a state of flux in the medico-legal domain than disease itself, evidence or malpractice. Current initiatives to improve it are encouraging and should be pursued. *Evidence, adverse effects*, and other topics in the law–health interface are increasingly better understood and defined in their substance and uses. This is necessary and must be lauded. For example, The Manitoba Evidence Act does so for evidence and other related terms in its court management.[104] Environmentalists in the United States point out the heterogeneity of terms like *adverse effect, risk, endanger*, or *threat* and their uses in U.S. law.[105]

The most crucial challenge relating to medical error at courts of law, at least in the eyes of an epidemiologist, is the confusion that may occur if medical error is considered synonymous with the harm that is imputed to it. It must be demonstrated that harm, the true reason for court litigation, is really the consequence and product of medical error. It must also be clear what the standard of care was and whether it was breached. Besides the error/harm confusion, a causal proof like showing that injury is really the consequence and product of error is another challenge.

Conclusions may be hampered by the poverty of epidemiological information about the particular error and its consequences. As an example, not only frequencies of error but also denominators for desired rates are imprecise, not available, or often unreported. Even less is known about the duration of the follow-up of error frequency and rates leading to limitations of our knowledge of error or harm incidence density as a product of error frequency and the duration of its follow-up. For example, the frequency of surgeons' errors and harm depends not only on the total number of surgeons under follow-up but also on the frequency and duration of surgical interventions they perform during a specific period of time. In most cases, we know little about the real occurrence (incidence, prevalence, duration, case fatality and other expressions of error frequency) that would be representative of a desired target population and valuable for a better knowledge of causal relationships.

Lawyers and courts do not necessarily know such intricacies and uses of epidemiology. They will require this information from medical expert witnesses and other physicians whether they represent the plaintiff or defendant. In important cases like class actions, lawyers can become very well versed in epidemiology and its value for the right verdict. Ideally, a common understanding between all litigating parties and court decision makers must be achieved before the trial itself to have a clear concept of the subject at hand.[106]

Cause–effect relationship demonstrations are a cardinal topic at courts of law. These demonstrations are crucial for any just verdict regarding causes of medical error, or medical error as cause of the harm for which the patient, their groups, or the community seek proper compensation.

Years of clinical experience on one side and judicial experience on the other are not guarantees of expertise in harm and injury problem solving at courts. Evidence-based clinical epidemiological reasoning, critical thinking, and decision making are increasingly necessary, important, and valued assets. Physicians, beware: men and women of law know more and more about domains that we have so far considered the exclusive hunting grounds of medicine and other health professions. No one likes to lose litigation or face.

References

1. Jenicek M. *Foundations of Evidence-Based Medicine.* Boca Raton, FL: Parthenon Publishing Group/CRC Press, 2003.
2. Clinton HR, Obama B. Making patient safety the centerpiece of medical liability reform. *N Engl J Med,* 2006;**354**(21):2205–8.
3. Horton R. The uses of error. *The Lancet,*1999;**353**(Feb 6):422–3.
4. Olivetti L, Fileni A, De Stefano F, Cazzulani A, Battaglia G, Pescarini L. The legal implications of error in radiology. *Radiol Med,* 2008;113:599–608.
5. Liang BA. Error in medicine: legal impediments to U.S. reform. *J Health Politics Policy Law,* 1999;**24**(1):27–58.
6. Gilmour JM. *Patient Safety, Medical Error and Tort Law: An International Comparison.* Final report. May, 2006. Available at: http://osgoode.yorku.ca/osgmedia.nsf/0/094676DE3FAD006A5852572, retrieved May 24, 2009.
7. Leflar RB, Iwata F. Medical error as reportable event, as tort, as crime: A transpacific comparison (Updated Version). *J Japan Law,* 2006; (**22**):39–76. Available at: http://criminology.law.usyd.edu.au/anjel/documents/ZJapanR/ZJapanR22/full%20article/ZJapanR22_07_Leflar_NEU%5B1%5D.2.pdf, retrieved May 25, 2009.
8. Chapter III. User Complaints. Pp. 13–31 in: Éditeur Officiel du Québec. An Act Respecting Health Services and Social Services. R.S.Q., chapter S-4.2, updated January 1, 2009. Available at: http://www.2.publicationsduquebec.gouv.qc.ca/dynamicSearch/telecharge.php?type=2&fi..., retrieved January 20, 2009.
9. Special Issue. *Evidence: Its Meanings in Health Care and in Law. J Health Politics Policy Law,* 2001;26 (Number 2, April).
10. Waller BN. *Critical Thinking. Consider the Verdict.* Englewood Cliffs, NJ: Prentice Hall,1988.
11. Jenicek M. Meta-analysis in medicine. Where are we and where we want to go. *J Clin Epidemiol,* 1989;**42**;35–44.
12. Jenicek M. *Méta-analyse en médecine. Évaluation et synthèse de l'information clinique et épidémiologique. (Meta-analysis in Medicine. Evaluation and Synthesis of Clinical and Epidemiological Information).* St-Hyacinthe et Paris: EDISEM et Maloine, 1987.
13. Brennan TA, Carter RF. Legal and scientific probability of causation of cancer and other environmental disease in individuals. *J Health Politics and Law,* 1985;**10**:33–80.

14. Black B, Lilienfeld DE. Epidemiologic proof in toxic tort litigation. *Fordham Law Review*, 1984;**52**:732–85.

15. Lilienfeld DE, Black B. The epidemiologist in court: some comments. *Am J Epidemiol*, 1986;**123**:961–4.

16. Teret SP. Litigating for the public's health. *AJPH*, 1986;**76**:1027–9.

17. Norman GR. Science, public policy, and media disease. *CMAJ*, 1986;**134**:719–20.

18. Cole P. Epidemiologist as an expert witness. *J Clin Epidemiol*, 1991;**44**(suppl 1):35S–39S.

19. Editorial. Greater use of expert panels proposed as additional means of presenting epidemiologic evidence to the courts. *Epidemiol Monitor*, 1989;**10**(4):1–3.

20. Holden C. Science in court. *Science*, 1989;**243**:1658–9.

21. Aldisert RJ. *Logic for Lawyers. A Guide to Clear Legal Thinking*. Third Edition. Notre Dame, IN: National Institute for Trial Advocacy (NITA), 1997.

22. Jenicek M, Hitchcock DL. *Evidence-Based Practice. Logic and Critical Thinking in Medicine*. Chicago, IL: American Medical Association Press, 2005.

23. Rothman KJ. Sleuthing in hospitals. *N Engl J Med*, 1985;**313**:258–9.

24. Istre GR, Gustafson TL, Baron RC, Martin DL, Orlowski JP. A mysterious cluster of deaths and cardiopulmonary arrests in a pediatric intensive care unit. *N Engl J Med*, 1985;**133**:205–11.

25. Buehler JW, Smith LF, Wallace EM, Heath Jr CW, Kusiak R, Herndon JL. Unexplained deaths in a children hospital. An epidemiologic assessment. *N Engl J Med*, 1985;**313**:211–6.

26. Sacks JJ, Stroup DF, Will ML, Harris EL, Israel E. A nurse-associated epidemic of cardiac arrests in an intensive care unit. *JAMA*, 1988;**259**:689–95.

27. Elsevier Science (USA). *Miller-Keane Encyclopedia and Dictionary of Medicine, Nursing, and Allied Health*. Edited by M O'Toole. 7th Edition. Philadelphia: Saunders, 2003.

28. Mohr JC. American medical malpractice litigation in historical perspective. *JAMA*, 2000;**283**(13):1731–7.

29. Gostin L. A public health approach to reduce error. Medical malpractice as a barrier. *JAMA*, 2000;**283**(13):1742–3.

30. Studdert DM, Mello MM, Gawande AA, Gandhi TK, Kachalia A, Yoon C, et al. Claims, errors and compensation payments in medical malpractice litigation. *N Engl J Med*, 2006;**354**(19):2024–33.

31. Reason J. *Human Error*. Cambridge, UK: Cambridge University Press, 1990.

32. Regenbogen SE, Greenberg CC, Studdert DM, Lipsitz SR, Zinner MJ, Gawande AA. Pattern of technical error among surgical malpractice claims. An analysis of strategies to prevent injury to surgical patients. *Ann Surg*, 2007;**246**(5):705–11.

33. Karl RC. The origins of malpractice claims. *Ann Surg*, 2007;**246**(5):712–3.

34. Rogers SO Jr., Gawande AA, Kwaan M, Puopolo AL, Yoon C, Brennan TA, Studdert DM. Analysis of surgical errors in closed malpractice claims at 4 liability insurers. *Surgery*, 2006;**140**(1):25–33.

35. Holder AR. Medical errors. *Hematology*, 2005;2005(Jan):503–6.

36. Clapp JE. *Random House Webster's Dictionary of Law*. New York: Random House, 2000.

37. Collin PH (Editor). *Dictionary of Law.* 2nd Edition. Chicago: Fitzroy Dearborn Publishers, 1999.
38. Conolly MJ, Nolan JR, Black HC. *Black's Law Dictionary with Pronunciations.* Abridged Fifth Edition. St. Paul, Minn: West Publishing Company, 1983.
39. Garner BA (Editor). *Black's Law Dictionary.* Seventh Edition. St. Paul, MN: West Group, 1999.
40. Dukelow DA. *The Dictionary of Canadian Law.* Third Edition. Scarborough, ON: Thomson Carswell, 2004.
41. Wikipedia, the Free Encyclopedia. *Medical Error.* Available at: http://en.wikipedia. org/wiki/Medical_error, retrieved December 23, 2008.
42. Dekker SWA. *Errors in our understanding of human error: the real lessons from aviation and healthcare.* Technical Report 2003-01. Lund: University School of Aviation, 2003.
43. *A Dictionary of Epidemiology.* Fifth edition. Edited for the International Epidemiological association by M Porta. JM Last, S Greenland. New York: Oxford University Press, 2009.
44. Runciman WB. Shared meanings; preferred terms and definitions for safety and quality concepts. *MJA,* **184**(10):S41–S43.
45. US Department of Veteran Affairs. National Center for Patient Safety. *Glossary of Patient Safety Terms.* Available at: http://www.va.gov/NCPS/glossary.html, retrieved January 14, 2009.
46. Davies JM, Hébert P, Hoffman C (The Royal College of Physicians and Surgeons of Canada). *The Canadian Patient Safety Dictionary.* Ottawa: Royal College of Physicians and Surgeons of Canada, October 2003. Available at: http://rcpsc.medical.org/publications/PatientSafetyDictionary_e.pdf -, retrieved November 8, 2008.
47. Croskerry P, Cosby KS, Schenkel SM, Wears RL. *Patient Safety in Emergency Medicine.* Philadelphia: Wolters Kluwer | Lippincott Williams & Wilkins, 2009.
48. World Alliance for Patient Safety (L. Leape and S. Abookire, Principal Authors). *WHO Draft Guidelines for Adverse Event Reporting and Learning Systems. From Information to Action.* Geneva: World Health Organization, Document WHO/ EIP/SPO/QPS/05.3, 2005.
49. Dukelow DA, Nuse B. *The Dictionary of Canadian Law.* Second Edition. Scarborough, ON: Carswell/Thomson Professional Publishing, 1995.
50. *Osborn's Concise Law Dictionary.* 10th Edition. Edited by M Woodley. London: Sweet & Maxwell, 2005.
51. *A Dictionary of Law.* Fifth Edition, reissued with new covers. Edited by EA Martin. Oxford, UK: Oxford University Press, 2003.
52. Jenicek M. *A Physician's Self-Paced Guide to Critical Thinking.* Chicago, IL: American Medical Association Press, 2006.
53. Glossary (pp. 259–278) in Reference 22.
54. Blackburn S. *The Oxford Dictionary of Philosophy.* Oxford: Oxford University Press, 1996.
55. Yogis JA, Cotter C. *Barron's Canadian Law Dictionary.* 6th Edition. Hauppauge, NY: Barron's Educational Series, Inc., 2009.
56. Gilmour J, Author of Reference 6: Personal communication.

57. Jenicek M. *Clinical Case Reporting in Evidence-Based Medicine.* Second Edition. London: Arnold and Oxford University Press, 2001.
58. The standard of care. Chapter 6, pp. 152–175 in: Merry A, McCall Smith A. *Errors, Medicine and the Law.* Cambridge, UK: Cambridge University Press, 2001.
59. Wikipedia, the Free Encyclopedia. *Standard of care.* Available at: http://en.wikipedia.org/wiki/Standard_of_care, retrieved August 26, 2009.
60. MedicineNet.com. *Definition of Standard of care.* Available at: http://www.medterms.com/script/main/art.asp?articlekey=33263, retrieved August 26, 2009.
61. Torrey T. for About.com. *Standard of Care. Definition.* Available at: http://patients.about.com/od/glossary/g/standardofcare.htm, retrieved August 26, 2009.
62. Haney C. Psychology and legal change: On the limits of a factual jurisprudence. *Law Human Behav,* 1980;**4**(3):147–99.
63. Haney C. Psychology and legal change. The impact of a decade. *Law Hum Behav,* 1993;**17**(4):371–98.
64. Roesch R. Creating change in the legal system. Contributions from community psychology. *Law Hum Behav,* 1995;**19**(4):325–43.
65. Dalby JT. On the witness stand. *Can Fam Physician (CFP),* 2007;**53**:65–70.
66. Sagall EL, Reed BC. *The Law and Clinical Medicine.* Philadelphia: J.B. Lippincott Company, 1970.
67. Picard EI. *Legal Liability of Doctors and Hospitals in Canada.* Toronto: The Carswell Company Limited, 1978.
68. Emson HE. *The Doctor and the Law. A Practical Guide for the Canadian Physician.* Second Edition. Toronto: Butterworths, 1989.
69. Sneiderman B, Irvine JC, Osborne PH. *Canadian Medical Law. An Introduction for Physicians, Nurses and Other Health Professionals.* Toronto: Thomson ⋅ Carswell, 2003.
70. Picard EI, Robertson GB. *Legal Liability of Doctors and Hospitals in Canada.* Fourth Edition. Toronto: Thomson ⋅ Carswell, 2007.
71. Wikipedia, the Free Encyclopedia. *Lawsuit.* Available at: http://en.wikipedia.org/wiki/Lawsuit, retrieved May 27, 2009.
72. Anon. *Stages of a Typical Civil Litigation.* Available at: http://www.samford.edu/schools/netlaw/dh2/casetutorial/Figure1.htm, retrieved May 27, 2009; http://samford.edu/schools/netlaw/dh2/casetutorial/Tortprocedures.html, retrieved May 27, 2009.
73. *The Civil Judicial Process.* Chapter 5, pp. 187–227 in: *The American Law Dictionary.* Edited by Renstrom PG. Santa Barbara, CA: ABC-CLIO, Inc. and Clio Press Ltd, 1991.
74. Hill AB. Observation and experiment. *N Engl J Med,* 1953;**248**:995–1001.
75. Hill AB. The environment and disease: Association or causation? *Proc Roy Soc Med,* 1965;**58**:295–300.
76. *Causal Inference.* Edited by KJ Rothman. Chestnut Hill: Epidemiology resources Inc., 1988.
77. Rothman KJ, Greenland S. Causation and causal inference, Chapter 15, pp 617–629 in: *Oxford Textbook of Public Health.* Third Edition. Edited by R Detels, WW Holland, J McEwen, GS Omenn. New York: Oxford University Press, 1997.
78. Evans AS. Causation and disease. A chronological journey. *Am J Epidemiol,* 1978;**108**:249–57.

79. *Clinic 2.* Assessment of causality. Pp. 11–19 in: Sagall EL, Reed BC. *The Law and Clinical Medicine.* Philadelphia: J.B. Lippincott Company, 1970.
80. Levin AS. Science in court. *Lancet*, 1987;**2**:1526.
81. Hoffman RE. The use of epidemiologic data in the courts. *Am J Epidemiol*, 1984;**120**:190–202.
82. Norman GR, Newhouse MT. Health effects of urea formaldehyde foam insulation: evidence of causation. *Can Med Ass J*, 1986;**134**:733–8.
83. Evans EA. Causation and disease: Effect of technology on postulates of causation. *Yale J Biol Med*, 1991;**64**:513–28.
84. Cole P, Goldman MB. Occupation. Pp. 167–184 in: *Persons at High Risk of Cancer.* Edited by JF Fraumeni Jr. New York: Academic Press, 1975.
85. Miller DW Jr., Miller CG. On evidence, medical and legal. *J Am Physic Surg*, 2005;**10**(3-Fall):70–5.
86. Miller DW Jr., Miller CG. *On Evidence, Medical and Legal.* Presented to the Seattle Surgical Society, May 22, 2006. Available at: http://www.donaldmiller.com/On%20Evidence%talk%20%to%20Seattle%20Surgical,pdf, retrieved December 27, 2007.
87. Aronson JK, Hauben M. Anecdotes that provide definitive evidence. *BMJ*, 2006;**333**(Dec 16):1267–9.
88. Kiene H, von Schön-Angerer T. Single-case causality assessment as a basis for clinical judgment. *Altern Ther*, 1998;**4**(1):41–7.
89. Kiene H. *Complementäre Methodenlehre der Klinischen Forschung. Cognition-based Medicine.* Berlin: Springer-Verlag, 2001.
90. Jenicek M. Clinical case reports and case series research in evaluating surgery. Part II. The content and form: Uses of single clinical case reports and case series research in surgical specialties. *Med Sci Monit*, 2008;**14**(10):RA149–RA162.
91. Rose G. Sick individuals and sick populations. *Int J Epidemiol*, 1985;**14**:32–8.
92. Savellos EE, Galvin RF. *Reasoning and the Law. The Elements.* Stamford, CT: Wadsworth/Thomson Learning, 2001.
93. Toulmin SE. *The Uses of Argument.* Updated Edition. Cambridge, UK: Cambridge University Press, 2003. (First published 1958).
94. Jenicek M. *Fallacy-Free Reasoning in Medicine. Improving Communication and Decision Making in Research and Practice.* Chicago, IL: American Medical Association Press, 2009.
95. Gelderman JM. Disclosure of error. *Ann Emerg Med*, 2006;**48**(5):631–2.
96. Moskop JC, Gelderman JM, Hobgood CD, Larkin GL. Emergency physicians and disclosure of medical errors. *Ann Emerg Med*, 2006;**48**(5):523–31.
97. Fein S, Hilborne L, Kagawa-Singer M, Spiritus E, Keenan C, Seymann G, et al. A conceptual model for disclosure of medical error. *Adv Patient Safety*, 2005;**2**:483–94. Available at: http://www.dtic.mil/cgi-bin/GetTRDoc?AD=ADA343207&location=U2&doc=GetTRDoc.^df, retrieved November 28, 2008.
98. AHRQ Morbidity & Mortality Rounds on the Web. *No Blood, Please. The Case and The Commentary.* Commentary by Brian A. Liang. Available at: http://www.webmm.ahrq.gov/printview.aspx?caseID=59, retrieved July 19, 2009.
99. AHRQ WEBM&M. *Case Archive.* A complete list of error and harm cases. Available at: http://www.webmm.ahrq.gov/caseArchive.aspx, retrieved July 19, 2009.

100. CTV.ca News Staff. *Top court upholds law that forced teen's transfusion.* June 26, 2009 update. Available at: http://www.ctv.ca/servlet/ArticleNews/print/CTVNews/20090626/supreme_court-09062..., retrieved June 26, 2009.

101. Nguyen AVT, Nguyen DA. *Learning from Medical Errors: Clinical Problems.* Oxford: Radcliffe Publishing, 2005.

102. Nguyen AVT, Nguyen DA. *Learning from Medical Errors: Legal Issues.* Oxford: Radcliffe Publishing, 2005.

103. Industrial Epidemiology Forum's Conference on Ethics in Epidemiology. Guest Editors WE Fayerweather, J Higginson, TL Beauchamp. *J Clin Epidemiol,* 1991;**44**:Supplement 1, suppl. Pp. 1–169.

104. *The Manitoba Evidence Act. C.C.S.M. c. E150.* Updated to January 16, 2009. Available at: http://wb2.gov.mb.ca/laws/statutes/cosm/e150.e.php, retrieved January 20, 2009.

105. Stansell K, Marvelli M, Wiener JB. *"Adverse Effects" and Similar Terms in U.S. Law.* A Report Prepared by the Duke Center for Environmental Solutions for the Dose Response Specialty Group of the Society for Risk Analysis, July 2005. Available at: http://www.sra.org/drsg/docs/Adverse_Effects_report.pdf, retrieved November 20, 2008.

106. Section 7.2. *Physicians in Courts of Law: Their Contributions to Decision-Making in Tort Litigation.* Pp. 229–141 in Ref. 22.

Conclusions

Thoughts to Think About

Essentially, all models are wrong but some are useful.

—George Edward Pelham Box, 1987

Experience is never at fault; it is only your judgment that is in error in promising itself such results from experience as are not caused by our experiments.

—Leonardo da Vinci, 1452–1519

Honorable errors do not count as failures in science, but as seeds for progress in the quintessential activity of correction.

—Stephen Jay Gould, 1998

No one wants to learn by mistakes, but we cannot learn enough from successes to go beyond the state of the art.

—Henry Petroski, 1992

It is almost as difficult to make a man unlearn his errors, as his knowledge. Mal-information is more hopeless than non-information: for error is always more busy that ignorance.

—Charles Caleb Colton, 1820

One way of dealing with errors is to have friends who are willing to spend the time necessary to carry out a critical examination.... An even better way is to have an enemy. An enemy is willing to devote a vast amount of time and brain power to ferreting our errors both large and small.... The trouble is that really capable enemies are scarce; most of them are ordinary.

—Georg von Békésy, 1899–1972

A book may be amusing with numerous errors, or it may be very dull without a single absurdity.

— Oliver Goldsmith (*The Vicar of Wakefield*, 1766), 1728–1774

A person who publishes a book appears wilfully in public with his pants down.

—Edna St. Vincent Millay, 1892–1950

Medical error and harm were, are, and will be with us despite our best abilities and competencies. It is up to us to do our utmost when dealing with such flaws in our well-intentioned endeavors.

We live in times when solid grounds have been built to study, prevent, and control medical error and harm. However, they are only grounds. Floors and roofs have yet to be built. We owe these foundations to enthusiastic health professionals and particularly to psychologists, ergonomists, operational research specialists, and informaticians who have all played an invaluable role in lathology. Adapting their critical contributions to the health domain is not enough. Health and disease care and management requires "insider" contributions, well beyond what we inherited from the fields of transportation, manufacturing, business, military arts, or policies and politics. It is only future work and understanding that will allow us to conclude what is more important in the Siamese twins represented by system and human medical error.

Medical error and harm domains still remain in a state of flux, developing and further refining their own identity, methodology, and subjects of attention. We do seem, nevertheless, to be improving.

Challenges in Focus

Fundamentally, more than a dozen major challenges remain in focus:

1. Confounding error and harm.
2. Persisting diversity of semantics and taxonomy.

3. Lack of epidemiology.
4. Dichotomy in lathology.
5. Lack of training in lathology.
6. Better knowledge, attitudes, and skills in the management of error and harm.
7. A need for better knowledge of causes of error and harm.
8. Challenge of communication.
9. Interaction between stakeholders in the error and harm domains in medicine.
10. Psychological, social, and legal challenges to perpetrators of error and creators of harm.
11. Material gains and losses related to error and harm.
12. Possible ethical challenges.
13. Individual human error versus system error.
14. Lack of pragmatic choices regarding what to do in lathology.
15. Unexplored roles, uses, and potentials of logic, critical thinking, and evidence in generating error management activities.
16. Legal considerations.

These problems merit some comments.

Confounding Error and Harm

The first challenge is that medical error is considered by many to be synonymous with medical harm. However, medical harm is not necessarily due to medical error in all instances. The causal relationship between error and harm needs to be better understood, and our decisions about them should be refined accordingly.

Persisting Diversity of Semantics and Taxonomy

The second challenge is the absence of unified, uniform, and standardized definitions and taxonomy and uses of them in national and international medical lathology worldwide. Adoptions, adaptations, and extrapolations are incumbent to any lathologist and must be specified for better uniformity and standardization to arise across the domain of medical error and harm. Perhaps further experience and diversity of research and practice objectives in the error and harm domain will show that sometimes various adaptations of semantics and taxonomy are not necessarily a bad thing.

Lack of Epidemiology

The third challenge is that medical error and harm are health problems like disease, handicaps, and levels of health and that related risks and benefits are subject to rules of fundamental, clinical, and field epidemiology. In the absence of better surveillance, notification, and the subsequent grasp of a more complete and representative picture of medical error and harm occurrence, we must work to do our best with whatever elements we have. Our knowledge of the real and representative magnitude of the problem remains incomplete, and cause–effect associations are built solely on what was more or less completely observed. Given this, our initiatives to prevent and control medical error may reach only the tip of a very asymmetrical iceberg.

Closer to us as concerned health professionals, the epidemiology of medical error and harm still remains rudimentary, perhaps suppressed for a moment by remarkable fundamental contributions to lathology coming from various domains of health and other sciences. This should not last long.

Dichotomy in Lathology

Fourth, not everything in the medical error and harm domain is equally tackled by epidemiology. As already stated, some are unique or infrequent; others are epidemic, pandemic, or endemic to the delight of epidemiologists. The former must initially be the subject of qualitative research and rigorous clinical epidemiological descriptions and analysis. Then, they must be presented and retained as units of observation for further error and harm case linkage and analysis that has already proved useful in the latter.

Unless demonstrated, structured, and methodologically developed otherwise, the domain of error and harm in health sciences will remain divided into two categories: (1) those of unique cases; and (2) those that occur in clusters or spread in numbers over time, in space, and in various types of individuals. As already mentioned in these pages, each category will often require different but complementary methodologies for medical error and harm problem solving.

Lack of Training in Lathology

The fifth challenge is a lack of general and uniform training and experience in lathology for health professionals both at the undergraduate and graduate levels.

Fortunately, an increasing number of university environments are introducing lathology at the core of their curricula.

Better Knowledge, Attitudes, and Skills in the Management of Error and Harm

The sixth challenge, related to the previous one, is to determine the required and necessary knowledge, attitudes and skills in lathology that will enable us to manage the medical error in the most rational way possible.

A Need for Better Knowledge of Causes of Error and Harm

The seventh challenge is our persisting need for better knowledge of causes of medical error. Without it, we can't build better-focused programs in medical error primary, secondary, and tertiary prevention and "freedom from medical error" promotion. So far, a great deal has been invested in the development of a methodology for detecting possible and definitely known causes, and much less has been invested in etiological research on unknown factors and markers of error and harm.

Challenge of Communication

The eighth challenge involves communication between all stakeholders in the medical error and harm problem. We all speak our own language, use our own vocabulary, and have our own ways of reasoning and decision making. Better harmony should supplant the cacophony currently evident in the area of medical error and harm.

Interaction between Stakeholders in the Error and Harm Domain in Medicine

The ninth challenge is the harmonious interaction or lack of between various national and international centers, institutes, agencies, and governmental bodies. So far, initiatives and accomplishments are rather fragmented in the medical error and harm domains—still more than in the more traditional domains of medicine and its specialties or surgery.

Psychological, Social, and Legal Challenges to Perpetrators of Error and Creators of Harm

The tenth challenge is a psychological one. Whatever we do, we act against human nature and the importance we place on learning and experience, goodwill, best ethics, and our quest for the best possible success of our decisions and actions. Our successes valorize us emotionally and advance us professionally and academically. The gratefulness of our patients stimulates to us to even better professional and human performance. However, medical error and harm are impeding our progress. When they happen, they undermine the efforts we have made to achieve the elusive ideal of human and professional perfection, but learning from them is invaluable. This does not mean then that we should simply ignore the latter to put the former in a better light.

Material Gains and Losses Related to Error and Harm

The eleventh challenge is the material gain or loss due to success or failure, often synonymous with medical error and harm. It means different things for all involved, namely, the physician, the patient, and other community stakeholders in the medical error challenge. This is perhaps the greatest obstacle facing better epidemiology and control of medical error. It is also an important element in legal matters related to medical error and harm for courts of law and interested parties.

Possible Ethical Challenges

The twelfth challenge leads to a question: What ethical problems exist in the domain of medical error and harm?

We may understand this better only later. In clinical trials, medical or surgical factors and their effects, adverse or otherwise, are under scrutiny. Their rules and the rules of their follow-up are set in advance. In studies of error and harm, as in most classical epidemiology, a noxious effect (harm) is of primary concern. It is more than an adverse effect! Rules of care in light of error are established most often a posteriori. Cost–benefit understanding of error and harm themselves and in relation to the desirable effect of beneficial factors still needs a more thorough analysis.

Individual Human Error versus System Error

The thirteenth challenge is that the debate regarding whether errors and harms are due to humans as individuals only or to the whole system (i.e., other individuals and technological and physical environments) continues. Our understanding today not only of individual professional competencies and dispositions but also of ways individual health professionals reason, argument (not argue!), and make decisions while using various kinds of evidence should further enhance our understanding of why and how health professionals succeed or fail.

Data and access to them through epidemiology should benefit from the 'no blame' strategy and attention to system factors. However, attention to human individual and group factors at the sharp (active) or blunt (latent) end should not be put on the backburner in causal research and measured at various levels of prevention.

Lack of Pragmatic Choices Regarding What to Do in Lathology

The fourteenth challenge is that we also need a pragmatic approach to choosing what is most worthy of our efforts in lathology. Whatever we consider necessary components of our endeavors may be worthy or unworthy for various further related activities. Figure 4.3 illustrated how each consideration of worthiness depends on previous components and on their worthiness in the chain and progress of our decision making.

The elements offered for evaluation will depend on the way cases, reports, and other information end up in our hands. Figure 4.3 outlines then the cascade of necessary components of good evidence for medical error and harm in light of its worthiness from one step to another. The quality of information on error and harm for evaluation purposes will depend on the quality of evidences (studies) serving such purposes.

Unexplored Roles, Uses, and Potentials of Logic, Critical Thinking, and Evidence in Generating Error Management Activities

The fifteenth challenge is that, from an epidemiologist's perspective, medical error and harm are health phenomena like any other, such as disease, health, elements of clinical and community care. Often, they have a much worse repute and connotation than disease itself. This does not, however, exclude them from rules of description of their occurrence, demonstration of their causes, or pragmatic choices and evaluation of their control and prevention.

Whether cases of error are unique and generally escape epidemiological attention or whether they occur in considerable numbers calling for some hard-core epidemiology and biostatistics, their formulations as topics of interest, analysis, evaluation, discussion, and implementation decisions are all ultimately exercises in logic and critical thinking supported by quantitative and qualitative methodologies.

Any evaluation, presentation of findings, and interpretation of them will after all then remain biostatistical and epidemiological considerations, an exercise in critical thinking and decision making. A summary of things to do and not to do at the time of critically evaluating what we have done in the error and harm domain and our conclusions and proposals to further manage the medical error and harm problem was presented as Table 4.3. It's worthy to look at it again in the spirit of our conclusions.

Legal Considerations

The sixteenth challenge is associated with the domain of health-related law. So far, lawyers and ethicists are much more interested in the clinical and medical domain of harm than in the domain of "pure" error. However, it is up to them and all health professionals to establish proofs that error led to harm.

The world of law will always be implicated in the harm domain and in determining the monetary, social, economical, and political impact of wrongdoings wherever they occur. Lawyers need to know the essentials of epidemiology, reasoning, and critical thinking applicable to health professionals. Common understanding of lathology problems remains fundamental in communication between the legal and medical worlds.

Moreover, medical error and harm will be seen, evaluated, and judged in the historical, cultural, and faith context of each society in which they occur. We need to acquire more understanding in these matters. Sensible directions will ensue.

Conclusions to the Conclusions

We live in exciting times for lathology. Many of those who are joining this movement may be dazzled and bewildered by the persisting Tower of Babel diversity and heterogeneity and by the overlaps in terminology, nosology, taxonomy, models, and resulting wide array of strategies and methodologies to improve and minimize the medical error and harm challenge. In the same spirit, readers of this book may feel somehow short-changed by the fact that we still do not have a uniform and standardized picture, language, and management of medical error across the planet. This result will be achieved soon, and these pages will simply be a modest contribution to such an ambitious goal.

Are you disappointed that this book does not offer unequivocal guidelines on how to deal with the medical error and harm problem? The aforementioned problems and challenges partly explain why this is so. Nevertheless, we have at least helped you think better in terms of lathology.

Please join us in our endeavor to advance the lathology from which our patients and communities will benefit the most and that will bring us the feeling that all our years of hard work, stress, ethos, striving for perfection, and devotion to Hippocratic ideals of the profession were not wasted. Such outcomes depend on all of us.

A Brief and (Hopefully) Harmonized Glossary

This glossary defines terms as they are used and understood in the spirit, context, and content of this book. The universal and official validity, in medicine and the health sciences, of the definitions presented here may be limited or vary. However, they represent, in our modest opinion, an optimal selection from human and medical error literature and experiences. How did we compile this glossary?

1. Two criteria were used for inclusion:
   a. Operational quality or adequacy for practical uses (inclusion/exclusion criteria, measurability, and taxonomy suitability).
   b. Compatibility with prevailing meaning and terminology in mainstream clinical epidemiology, evidence-based medicine and uses of evidence in reasoning, critical thinking, understanding and decision-making in medicine and allied health sciences.
2. Additional attention was paid to the overlapping of various definitions (like between an incident and error or adverse event). Such instances were avoided as much as possible.
3. Even some basic epidemiological, logic, critical thinking, evaluation, and medical terms were excluded since they would have unnecessarily encumbered the glossary. Only definitions perhaps less well known to health professionals at large, and more specific to lathology, were retained in this selection.
4. The terminology of law pertaining to error and harm is detailed in Chapter 8.
5. Whenever more than one definition is quoted, it is because of the definitions' complementarity and convergence rather than their differences and ensuing controversy.
6. Anything else besides the aforementioned should also help readers who may not be health professionals to consult these pages with ease.

Accident (in general): A specific, identifiable, unexpected, unusual, and unintended event that occurs in a particular time and place, without apparent or deliberate cause, but with marked effects. It implies a generally negative probabilistic outcome that may have been avoided or prevented had circumstances leading up to the accident been recognized and acted upon prior to its occurrence.[1] It is an event that was neither planned nor expected.

Accident (in medicine): (a) An adverse outcome not caused by chance or fate. Most accidents and their contributing factors are predictable, and the probability of their occurrence may be reduced through system improvements.[2] (b) An unexpected, specific, identifiable, unusual, and unintended (unplanned) event that occurs in a particular time and place of medical care, originally without an apparent or deliberate cause but with marked effects.

Action-based error (slip, in medicine): Attentional or perceptual failure of an action[3] (e.g., dispensing an excessively elevated dose of medication).

Active error (human, medical): Error or failure resulting from human behavior[4,5] made by a health professional who, in a health establishment, provides direct clinical or community care, acts, or services (e.g., operating surgeon, prescribing internist, consulting psychiatrist, nurse at floors or in the operating room); may be knowledge-based, rule-based, or skill-based.

Active failure (also patent failure, salient failure, proximate failure): (a) An event, action, or process that is undertaken, or takes place, during the provision of direct patient care and fails to achieve its expected claims. While some active failures may contribute to patient injury, not all do.[2] (b) Failure of reasoning, judgment, and ensuing decision made by the operator—that is, the, executing health professional responsible for directly delivering healthcare to its recipient (patient or community). (c) A failure that immediately precedes error and its consequences. (N.B. What is immediate, and what is not?)

Adverse drug reaction: (a) An undesirable effect expected or not, produced by the use of medication in a manner recommended or not. Its mechanisms may be known or still unknown given its newness. (b) A medication-related adverse event.[6]

Adverse effect: An unfavorable, undesirable, and harmful result of a correct process or action.[7-modified]

Adverse event (in one of the following ways): (a) An unexpected and undesired incident directly associated with the care and services provided to the patient.[2] (b) An incident that occurs during the process of providing healthcare and that results in patient injury or death. (N.B. See incident and critical incident definitions.)[2] (c) An adverse outcome for a patient, including injury or complication.[2] (d) An untoward incident, therapeutic misadventure, iatrogenic injury, or other adverse occurrence directly associated with care or services provided within the jurisdiction of a medical center, outpatient clinic, or other (health) facility.[8] (e) An injury related to medical management, in contrast to complications of disease. ... medical management includes all aspects of care ... Adverse events may be preventable or non-preventable.[6]

Argumentation (in general): Methodological employment or presentation of arguments.

Argumentation (medical): Methodological use or presentation of medical arguments to solve a problem in medical practice or research.

Argument in general (as vehicle of reasoning): A connected series of statements or reasons intended to establish a position — that is, leading to another statement as a conclusion. (A position or conclusion may be the diagnosis, patient admission or treatment decision or that an error and/ or harm occurred, etc.)

Argument (in medicine): A connected series of statements originating from a lived situation, experience, or research in medicine intended to establish a position (another statement as conclusion) in medical problem solving, understanding and decision-making.

Argument-based error: Misusing or omitting valid argument components (grounds, backing, warrant, qualifier, rebuttals, conclusions or claims) or using inappropriate arguments and linking them poorly in medical decision-making.

Backing (in modern argument): Body of experience and evidence supporting the warrant and grounds in modern argument and argumentation.

Blunt end: Layers of healthcare that precede direct contact with patients where error-related events occur at a site of latent error: health policy setters, managers, technology designers, etc. An error in programming an intravenous pump is a sharp end (see this entry) or active error. Designing and manufacturing a defective pump is a blunt end (latent) error.[9-modified]

Case series study (clinical): A detailed description and analysis of series of cases that explains the dynamics, pathology, management and/or outcome of a given disease (or of harm or error in our context).[10]

Case study: An in-depth examination, analysis, and interpretation of a single instance or event such as the occurrence of error or harm.

Causal tree (in root cause analysis): (a) A linear sequence of events established to trace flaws in the process or healthcare or practice. (b) An investigation and analysis technique used to record and display, in a logical, tree-structured hierarchy, all the actions and conditions that were necessary and sufficient for a given sequence to have occurred.[11]

Cause: In the error domain: An antecedent set of actions, circumstances or conditions that produce an event, or phenomenon (i.e. error and its consequences).[2]

In general: An event without which some subsequent event would not have occurred or because of which it occurred. An agent or act that produces some phenomenon (the effect).[12]

At courts of law: Each separate antecedent of an event. That which in some manner is accountable for a condition that brings about an effect or that produces a cause for the resultant action or state.[12]

Claim (in modern argument): Conclusion of the argument; proposition as a result of reasoning based on supporting grounds and inference from them.

Clinical care: See healthcare.

Clinical epidemiology: (a) The application of epidemiological knowledge, reasoning, and methods to study clinical issues and improve medical and other decisions when dealing with individual patients and groups of patients and to improve overall clinical care (N.B. and its outcomes).[13-modified] (b) Using the experience acquired in groups in reasoning and decision-making in the care of individuals.[14]

Clinical case report: A structured form of scientific and professional communication normally focused on an unusual single event (patient or clinical situation) to provide a better understanding of a case and of its effects on improved clinical decision-making.[10]

Clinimetrics: The domain concerned with qualification, counting, measurement and categorization of clinical observations, data and information.

Close call: An event or situation that did not produce patient injury, but which might have, given the patient's condition, a lack of rapid corrective intervention by the care provider or other circumstances. See also *near miss incident.*[9-modified]

Cognition: (a) Human faculty of processing information, creating, changing, and applying knowledge and preferences. (b) Mental functions and processes (thought, comprehension, inference, decision-making, planning and learning, abstraction, generalization, concretization, specialization, meta-reasoning, beliefs, desires, knowledge, preferences, intentions of individuals, objects, agents, and systems)[15,16-modified] in view of the development of knowledge and concepts culminating in both thought and action.

Cognitive error (bias): Pattern of deviation in judgment that occurs in particular (in our context) in medical and clinical situations or in medical research reasoning and conclusions.

Cognitive science: An amalgamation of disciplines including artificial intelligence, neuroscience, philosophy, and psychology. Within cognitive science, cognitive psychology is an umbrella discipline for those interested in cognitive activities such as perception, learning, memory, language, concept formation, problem solving, and thinking.[17]

Community medicine: Practice of medicine at the community (beyond individual) level and setting, having as a purpose to identify health problems and needs, and the means by which these needs may be met, and to evaluate the extent to which health services meet these needs.[13-modified]

Complication: A disease or injury resulting from another disease, health state and/or healthcare intervention related to them.[2-modified]

Contributing factor: (a) Any factor that increases the risk of error besides the factors considered to be a principal cause. (N.B. Nowhere did we find a clear distinction between what is a principal cause and a contributing factor.) (b) An external, organizational, staff and/or patient-related circumstance, action or influence thought to have played a part in the origin or development of an incident or to have increased the risk of an incident.[18]

Critical incident: An incident leading to or resulting in serious harm (loss of life, limb, or vital organ) to the patient, or the significant risk thereof, requiring an immediate investigation and response, and subsequent actions to reduce the likelihood of recurrence.[2-modified]

Critical thinking: The intellectually disciplined process of actively, skillfully conceptualizing, applying, synthesizing, or evaluating information gathered from, or generated by, observation, experience, reflection, reasoning, or communication as a guide to belief or action.[19] Reasonable reflective thinking that is focused on deciding what to do or believe.[20]

Effectiveness/efficiency/efficacy triad: Evaluation of the impact of medical, clinical or community interventions (error and harm control included) in habitual conditions (**effectiveness**), ideal conditions (**efficacy**) and in relation to the cost (material and human means) invested in health interventions (**efficiency**).[14]

Epidemiology: (a) The study of the occurrence and distribution of health-related states or events in specified populations, including the study of determinants influencing such states, and the applications of such knowledge to control health problems.[13] (b) Epidemiology may be subdivided into clinical epidemiology, field epidemiology, and fundamental epidemiology. (See separate entries for these terms.)

Error (in general): An act of commission (doing something wrong) or omission (failing to do the right thing) that leads to an undesirable outcome or significant potential for such an outcome.[9]

Error (medical): An inaccurate or incomplete assessment of a patient's risks and diagnosis, conservative or radical treatment, prognosis, follow-up and care, including disease, injury, syndrome, behavior, infection and other subjects of care.[21] ... It usually reflects a deficiency in the system of care.[6]

Error chain: Sequence of latent and active errors and other events leading to a harming outcome. Root cause analysis helps in the understanding of an error chain.

Error creators: Latent error-generating individuals at the 'blunt end': New medical technologies engineers, developers of clinical guidelines and work protocols, organizers of clinical care.

Error outcomes: Incidents, accidents and their physical, mental and social consequences for the patient, health professional, and community. They may be negative (i.e. causing harm) or, less frequently, positive (changing health for the better).

Evaluation (in medical lathology): (a) A process that attempts to determine as systematically and objectively as possible the relevance, effectiveness and other characteristics of activities, programs and policies to prevent and control medical error and its consequences as well as the impact of such activities, programs, and policies in light of their objectives.[13-modified] (b) The systematic assessment of the activity (medical or other) and/or the outcomes of such an activity, program, or policy, compared to a set of implicit or explicit standards as a means of contributing to the improvement of the activity.

Event (medical): (a) Something that happens to or with a patient and/or a health professional or any other person involved in healthcare. (b) In this context (as in the case of 'error'), anything that leads to injury of the patient or poses a risk of harm.[6]

Evidence (in law, in relation to medicine): Any kind of proof of facts at courts: *real evidence* (defective medical instruments), *direct evidence* (photographs, prescriptions, patient records, oral testimonies), *circumstantial evidence* (circumstances in which the medical error occurred), *hearsay* (third party information quoted to the health professional), *confession* by the defendant, *dying declarations* (made prior to death by a patient), *other statements that prove to be relevant at the hearing* (records of some routine observations), *declarations against interest* (information showing the motive of a health professional's actions), *connected events proving design or intent* (repeated past events contradicting an isolated happening), *evidence as to the character of the defendant* (health professional or patient). Their admissibility is decided by the court. The term evidence in this context has a broader meaning than in medicine.[22]

Evidence (in medicine): Any data or information, whether solid or weak, obtained through experience, observational research or experimental work (trials) and their synthesis.

Evidence-based medicine: Practice of medicine based on the integration of best research evidence with clinical expertise and patient values.

Evidence-based error: Using poor or inappropriate evidence in argumentation and decision-making in clinical practice and research.

Failure: Non-performance or inability of the system or component (human included) to perform an intended function in a specific person/time/place context and specific conditions to reach an intended objective. Therefore, not all faults are failures.

Fallacy (in general): (a) A mistake or flaw in reasoning and argument. (b) A violation of the norms of good reasoning, rules of critical discussion, dispute resolution, and adequate communication.[23]

Fallacy (in medicine): Any error in reasoning pertaining to a health problem and its supporting evidence(s) and pertaining to the handling of evidence in our reasoning, and throughout the process of argumentation interfering with the best possible understanding and decision making in the task of health problem solving.[24]

Failure mode: (a) Any manifestation of error such as a precocious change in care, stopping or replacing ongoing care or a reporting decision in this sense. (b) Any error or defect in a process, design, or item, especially those that affect the customer (N.B. patient, in our case), and that can be potential or actual.[25] (c) The characteristic manner in which a failure occurs. Within a failure mode diagnostic model, failure modes represent specific ways in which a system, device, or process can fail.[26] (d) The physical or functional manifestation of a failure. For example, a system in failure mode may be characterized by a slow operation, change of operation, addition of a new operation, incorrect outputs, or incomplete (or precociously completed) termination of execution. Any surgical, obstetrical or gynaecologic operation fits the activity as quoted above.[27-modified]

Failure mode and effect analysis (FMEA): (a) A procedure for analysis of potential failure modes, i.e., descriptions of the way the failure occurs and the manner by which a failure is observed.[28] (b) A step-by-step approach for reviewing and identifying all possible failures in organizational process, policies, and procedures, design, manufacturing or assembly process, or product or service.[29]

Field epidemiology: (a) The practice of epidemiology in the community, commonly in a public health service.[13] (N.B. Sometimes termed 'shoe-leather epidemiology' given its on-foot door-to-door practice of inquiry and services delivery.) (b) Using experience gathered from individuals when dealing with health and disease at the community level.[14]

Flaw: Any characteristic, deficiency, or inadequacy in technology, operation, human reasoning and decision-making contrary to expectations from the expected function, task, and outcomes of a healthcare event. This term denotes any variable that may be considered a possible (often yet to be proven) causal factor or marker related to domains of risk and prognosis in the domain of medical error and its consequences.

Forcing function: An aspect of the design of a clinical tool or procedure that prevents a potentially harming target action from being performed or that allows its performance only if another specific action is performed first. Examples: Security caps on medications, authorization and supervision of medical care actions.

Fundamental epidemiology: Development and testing of basic manners of reasoning and decision-making in epidemiology through various domains such as philosophy, epistemology, logic and critical thinking, quantitative methods (biostatistics), and/or qualitative research in dealing with health and disease at the group of individuals and community levels.

Grounds (in modern argument): Specific facts, data, and information from research and/or practice as a basis for a claim.

Harm (in general): Injury or damage to people, property, or environment.[30]

Harm (medical): A temporary or permanent physical impairment in body functions (including sensory functions, pain, disease, injury, disability, death) and structures, as well as suffering and other deleterious effects due to a disruption of the patient's mental and social well-being.[31-33; reworded] Unintended physical injury resulting from or contributed to by medical care (including the absence of indicated medical treatment), that requires additional monitoring, treatment, or hospitalization, or that results in death. Such injury is considered medical harm whether or not it occurred within the hospital. … Some errors do indeed result in medical harm, but many errors do not; conversely, many incidents of medical harm are not the result of any errors.[34]

Hazard: A set of circumstances or a situation that could harm a person's interests, such as their health or welfare.[2] In epidemiology, the inherent capability of an agent or a situation to have an adverse effect.[13] 'Risk' is the probability of disease occurrence, in terms of incidence or mortality densities, whereas 'hazard' refers to probability in the field of prognosis, usually as a function of mortality, although any outcome other than death can also apply.[14] An error capable of causing harm; a potential source of harm.[30] Any threat to safety, e.g., unsafe practices, conduct, equipment, labels, names.[6]

Healthcare: (a) Services provided to individuals or communities to promote, maintain, monitor, or restore health, including self-care.[7] Those services may be provided by physicians (medical care), all other health professionals (clinical care or community care) and/or civic bodies (in public health) either in hospitals, clinics, medical/health offices, or communities at large. (b) Services provided to individuals or communities by agents of the health services or professionals to promote, maintain, monitor or restore health. Healthcare is not limited to medical care, which implies action by or under supervision of a physician. The term is sometimes extended to include health-related self-care.[13]

Healthcare failure mode and effect analysis (HFMEA): (a) Failure mode and effect analysis (FMEA) adapted to medicine and clinical care. (b) A systematic, prospective, multidisciplinary team-based risk analysis that identifies and assesses the effect of potential errors or system failures.[35]

Hindsight: A way of reasoning and drawing conclusions about an event after it has happened.

Hindsight bias: The tendency to judge events leading up to an accident as errors because the bad outcome is known; the more severe the outcome, the more likely that decisions leading up to this outcome will be considered errors. Judging the antecedent decisions as errors implies that the outcome was preventable.[9]

Human error (in general): (a) The failure to complete a planned action as it was intended, or when an incorrect plan is used in an attempt to achieve a given aim.[2] (b) All those occasions in which a planned sequence of mental and physical activities fails to achieve an intended outcome, and when these failures cannot be attributed to the intervention of some chance agency.[36] (c) Failures committed either by an individual or a team of individuals.

Human error (in medicine): Operator's error in reasoning, understanding and decision-making about the solution of the health problem and/or in the ensuing sensory and physical execution of a task in clinical and/or community care.

Iatrogenic illness: Adverse effect of preventive, diagnostic, therapeutic, surgical, and other medical, sanitary and health procedure(s), interventions, or programs attributable as resulting from the activity of a health professional.[13] It also includes other harmful occurrences that were not natural occurrences of the patient's disease. It does not automatically imply the culpability or responsibility of the physician or hospital, or that the illness was necessarily preventable.[37]

Incident (in general): Event, process, practice, or outcome that is noteworthy by virtue of the hazards it creates for, or the harm it causes, subjects. All accidents are incidents, but not all incidents are accidents.[38,39-modified]

Incident (in medicine): (a) Event, process, practice or outcome that is noteworthy by virtue of the hazards it creates for, or the harm it causes, patients.[2] (b) An event or circumstance that could have resulted, or did result in unintended or unnecessary harm to a person and/or in a complaint, loss or damage.[7] (c) Adverse events, critical incidents, sentinel events and near misses in clinical and community care. (N.B. See separate entries for these terms in this glossary.) Incident means any of them.[40]

Incident reporting: Identification of occurrences that could have led, or did lead, to an undesirable outcome.[9] Incident reporting is a part of the epidemiological surveillance of medical error and harm.

Lapse: Failures of memory that do not necessarily manifest themselves in actual behavior and may only be apparent to the person who experiences them.[36]

Lapse (in medicine): A memory-based error, a memory generated failure. (Example: Forgetting a patient's allergy to an antibiotic.)

Latent error (medical) or latent failure: (a) Error that results from underlying system failure(s).[4,5] (b) Less apparent failure(s) of organization or design that contributed to the occurrence of errors or allowed them to cause harm to patients. [9,36] (c) Any part in a chain of flawed more or less remote events that precede (and may lead to) the active error and harmful action. (d) A defect in design. Organization, training or maintenance in the system that leads to operator errors and whose effects are typically delayed.[6] (e) Latent error or failure may then be human, technical, external, and/or design-construction-material related.

Latent or remote event (cause, event, condition): (a) A factor that, in the past, might have contributed to a predisposition to the event, effect, or phenomenon such as error.[2-modified] (b) A "blunt end" event.

Lathology: The domain of the study and management of error and harm (medical error and harm in the context of this book). From the Greek *lathos*, i.e., error, and *logos*, i.e., study.

Learning curve: The description of the development and acquisition of a new surgical or medical skill in the search for moments, levels or stages of reaching a potential for lower-than-expected success rates or higher-than-expected error, complication, and/or harm rates. Learning curve study and analysis is used increasingly in health professional training progress evaluation.

Malpractice: Improper or unskillful conduct on the part of a medical practitioner that results in injury to the patient … It generally describes professional misconduct or negligence on the part of a person delivering professional services.[41]

Medical care: See healthcare.

Medical error: Human or system error (as defined already in this book) in healthcare and community medicine and medicine-related public health. Unintentionally being wrong in conduct or judgment in medical care.[7-modified]

Medical harm: See **Harm (medical)**

Medication error: (a) Human error as defined in the process of providing medications to patients.[2] It may be due to an omission, an incorrect dose, a failure to order a particular procedure, an incorrect form, an incorrect time, an incorrect route, a deteriorated drug, an incorrect rate of administration, an incorrect administration technique, or an incorrect dose preparation related to the dispensing a pharmacological agent (drug).[42-modified] (b) Any preventable event that may cause or lead to inappropriate medication use or patient harm when the medication is in the control of the health professional, patient, or consumer.[43]

Meta-cognition: Reflection about the thought processes that led to a particular diagnosis or other clinical or scientific decisions in order to consider whether biases or cognitive shortcuts may have had a detrimental effect.[9-modified]

Meta-evaluation: Evaluation of evaluations.

Mistake (in general): Deficiency or failure in the judgmental and/or inferential processes involved in the selection of an objective or in the specification of means to achieve it, irrespective of whether or not the actions directed by this decision-scheme run according to plan.[36]

Mistake (medical): (a) Incorrect reasoning, problem solving, and/or decision-making or sensory-motor action in clinical care. (b) Inappropriate planning of knowledge-based and rule-based actions resulting in errors in clinical care or in medical research.[41,5-modified] (c) Commission or omission with potentially negative consequences for the patient that would be judged wrong by skilled and knowledgeable peers at the time it occurred, independent of whether there were any negative consequences.[44]

Mitigating factor: An action or circumstance that prevents or moderates the progression of an incident towards harming the patient.[4] (N.B. Mitigating factors have a different meaning in law.)

Near miss: (a) An incident that did not cause harm.[7] (b) Any event that could have had adverse consequences, but did not; an event having similar characteristics as harm (adverse effect) producing with the exception of its harm/adverse effect outcome(s).

Node (in root cause and failure mode and effect analyses): Events considered important and of particular interest as connecting points in the study of cause-effect relationships.

Operators: "Sharp-end," "acting" error makers like operators of machines in manufacturing or drivers in transportation. In health, direct clinical care providers (prescribing physicians, operating surgeons, care dispensing psychiatrists) using mental and sensorimotor-designed tools, machines, instruments, and materials (drugs, replacement substances).

Outcome: (a) Product, result or practical effect of healthcare in terms of patient health and associated costs.[2-modified] (b) "What happens" when acting on patient (or community) health and disease. (c) The health status of an individual, a group of people or a population that is wholly or partially attributable to an action, agent, or circumstance.[7]

Patient: In the context of medical lathology, individuals, groups, populations or entire communities that require and/or receive healthcare provided by healthcare professionals.[45]

Patient safety: The reduction and mitigation of unsafe acts (N.B. and of their undesirable consequences) within the healthcare system, as well as through the use of best practices shown to lead to optimal patient outcomes.[2]

Prevention: Actions that control disease occurrence either by preventing new cases (**primary** prevention), lowering disease prevalence, usually by shortening the duration of cases (**secondary** prevention), minimizing the severity and sequellae of cases beyond primary and secondary control (**tertiary** prevention), or actions to limit iatrogenesis (**quaternary** prevention).[13-reworded]

Probabilistic risk analysis: A probability-based description and analysis of possible associations between adverse events and their potential causes.

Process (in healthcare evaluation): (a) A course of action or sequence of steps, including what is done and how it is done.[2] (b) A "how does it work" characteristic of an activity or system within which errors occur. (c) All that is done to patients in terms of diagnosis. Treatment, monitoring, and counseling.[9]

Proposition (in critical thinking): An assertion (affirmation or denial) of something which is capable of being judged true or false. (Ex. Diagnosis. An error is the cause of harm in this patient.)

Public health: Organized effort, structures, policies and programs by society to protect, promote, restore, and improve health and prolong the life of the community.[13-modified]

Qualifier (in modern argument): Strength (degree of certainty) conferred by the warrant on the inference from grounds to claim.

Quality of healthcare: Degree to which health services for individuals and populations increase the likelihood of desired health outcomes and are consistent with current professional knowledge.[9] The extent to which a healthcare service or product produces a desired outcome or outcomes.[7]

Qualitative research (in error and harm domain): Any kind of research on error and harm that produces findings not arrived at by means of statistical procedures or other means of quantification about persons' lived experience with error and harm, behavior facing them, but also on organizational, functional, social or interactional relationships between health professionals and their patients and community. [46-modified]

Rebuttals (in modern argument): Exceptional (exclusionary) circumstances undermining or weakening or nullifying final conclusions (claim) of an argument.

Risk: (a) The probability of danger, loss, or injury within the healthcare system.[2] (b) In epidemiology: The probability that an event will occur, e.g. that an individual will become ill or die within a stated period of time or by a certain age.[16,30] (N.B. Usually, an unfavorable outcome or event is in focus, but probabilities of other events [even beneficial ones] may be quantified as "risk.")

Risk management: Organizational activities in healthcare designed to prevent patient injury or moderate actual financial losses following an adverse outcome.[2]

Root cause: A causal factor that, if corrected, would prevent recurrence of the incident, encompassing and derived from several contributing causes such as system deficiencies, management failures, performance errors and inadequate organizational communication.[47]

Root cause analysis: (a) A structured retrospective and already acquired knowledge-based process for identifying the causal or contributing factors underlying adverse events or other critical incidents.[9-modified] Primarily of descriptive value. (b) A systematic iterative process whereby the factors that contribute to an incident are identified by reconstructing the sequence of events and repeatedly asking "why?" until the underlying initial ("root") causes have been elucidated.[29] (c) An analytic tool that can be used to perform a comprehensive, system-based review of critical incidents. It includes the identification of root and contributory factors, identification of risk reduction strategies, and development of action plans along with measurement strategies to evaluate the effectiveness of the plans.[38]

Rule-based error: Error due to the misapplication of or the failure to apply a good rule, or a bad rule.

Sentinel event: (a) An adverse event in which death or serious harm to a patient or community has occurred.[9-modified] All sentinel events are not expected and they may be unacceptable such as wrong side or wrong organ operations. (b) Egregiousness of the harm and the likelihood that its investigation will reveal serious problems in current policies and procedures.[9-modified] (c) Unexpected incident related to the system or process deficiencies and/or human error, which leads to death or major and enduring loss of function (and/or anatomy) for a recipient of healthcare services.[40]

Sharp end (site or event): Health professional(s) or parts of the health system and care in direct contact with patients in medical error and harm situations and events.

Side effect (in general): (a) An effect, other than that intended, produced by an agent. (See also 'adverse effect or reaction').[7] (b) Any effect that occurs besides the main, expected, and desired one.

Side effect (in medicine): An effect, other than that intended, produced by a preventive, diagnostic, or therapeutic procedure or regimen. Not necessarily harmful.[13]

Significant event analysis: A qualitative method of clinical audit based on a synthesis of traditional case reviews and research principles of the critical incident inquiry to perform a more in-depth, structured analysis of an event identified as 'significant' by a healthcare team.[48]

Skill-based errors: Slips and lapses; errors in execution of correctly planned actions, encompassing both action-based errors (slips) and memory-based errors (lapses).[3]

Slip (in general): An action not in accord with the actor's intention, the result of a good plan, but a poor execution,[49] hence physical, sensory/motor failure.[50]

Slip (medical): (a) Inappropriate action and execution, incorrect execution of a correct action sequence,[51,36-modified] due also to competing sensory or emotional distractions, fatigue, or stress.[9] (b) Failure in schematic (often learned and mastered) behavior due to distractions rather than to professional qualification and experience. May be related to execution (goals, intention, action specification, execution itself) and/or evaluation (perception, interpretation, action evaluation).[51] (c) Action-based error based on attentional or perceptual failure of an action, e.g. dispensing an elevated dose of medication. See 'technical slip'.

Structure (of health activity or system): (a) Supporting framework of essential parts including all elements of the healthcare system that exist before any actions or activities take place.[2] (b) A "consisting of what and how is it organized" characteristic of an activity or system within which errors occur. (c) The setting in which care occurs and the capacity of that setting to produce quality. ... Structural measures such as credentials, patient volume, and academic affiliation.[9]

Standard of care: A set of steps that would be followed or an outcome that would be expected.[2] ... A level of measure, rather than a rule or policy.[2]

"Structure-Process-Outcome" triad: Triangular evaluation of the quality of clinical and community care. Such a view of quality depends on what is part of care, how the care is delivered and what is its impact, i.e. favorable outcomes in patients or members of the community.[52]

System (in general): A set of interacting and independent entities, real or abstract, forming an integrated whole. ... A set of relationships that are differentiated from relationships of the set to other elements, and from relationships between an element of the set and elements not part of the relational regime. ... Systems have a structure that is defined by its parts and their composition. They have behavior that involves inputs, processing, and outputs of material, information, or energy. ... Various parts of a system have functional as well as structural relationships between each other.[53]

System (in healthcare): A set of interdependent components interacting to achieve a common claim. ... System characteristics include complexity and coupling.[2]

System analysis: Analysis of a broad system of events with a chain of contributing factors combined in the future with incident analysis and with anticipation of future problems based on current experience.[54]

System approach: Alternative, complementary, or expanded approach to errors and harm attributable to health professionals as individuals, analyzing systems (rather than the behavior of individuals only) in order to identify situations or factors likely to give rise to human error and harm and to implement "system" changes as remedial measures, and to reduce the occurrence of errors and harms and minimize their impact on patients.[9-modified]

System error (in medicine): Error imputable to technology and end environment of medical care and its interaction with their users, i.e., health professionals, as operators of the system and their recipients, i.e., patients and other receivers of therapeutic and preventive care.

System failure: A fault, breakdown, or dysfunction within an organization's operational methods, processes and infrastructure.[7]

Taxonomy: Classification of entities of interest. Usually it covers a particular domain like error and/or harm in our case.

Team error: A human error committed by a group of individuals within their interaction while planning, performing, and evaluating a task.[15]

Technical slip in medicine (examples): Wrong writing of orders, inappropriate clinical maneuver execution, dispensing technique, recording, and information storage. (See the 'slip' entry for basic definitions of that term.)

Thought experiment: A proposal for an experiment that would test a hypothesis or theory, but cannot actually be performed due to practical, ethical or other limitations (cultural, value, etc.); instead its purpose is to explore the potential consequences of the principle in question.[55-modified] A device of the imagination used to investigate the nature of things.[56]

Time-to-event analysis: An analysis of intervals between two events and the meaning of such intervals in relation to the nature of events and relationships between them. Used in clinical trials, prognosis, survival studies, lathology, and elsewhere. In the broadest sense, it includes also analyses of incubation periods of diseases and generation time in the infectious disease domain.

Tort (in law): A wrongful act, other than a breach of contract, that results in injury to another person, property, reputation, or some other legally protected right of interest, and for which the injured party is entitled to a remedy of law, usually in the form of damages.[57]

Unsafe act: Any events that result from a criminal act or act done under the effects of any substance abuse impairing the health provider. Errors, violations, and sabotage (see separate definitions in this glossary) are all unsafe acts.[2]

Violation: Deviation from safe operating practices, procedures, standards, or rules.[36,58]

Warrant (in modern argument): General rule to infer a claim from grounds.

References

1. Wikipedia, the free encyclopedia. *Accident.* 2 pages at http://en.wikipedia/org/wiki/accident, retrieved November 11, 2008.
2. Davies JM, Hébert P, Hoffman C. *The Canadian Patient Safety Dictionary.* Ottawa: Royal College of Physicians and Surgeons of Canada, October 2003. 58 pdf pages at http://rcpsc.medical.org/publications/PatientSafetyDictionary_c.pdf, retrieved January 21, 2009.
3. Ferner RE, Aronson JK. Clarification on terminology in medication errors. Definitions and classification. *Drug Safety,* 2006;**29**(1):1011–22.
4. White JL. *Adverse Event Reporting and Learning Systems: A Review of the Relevant Literature, June 25, 2007.* A Paper Prepared for CPSI. Edmonton and Ottawa: Canadian Patient Safety Institute (CPSI), 2007. 80 pdf pages at www.patientsafetyinstitute.ca/uploadedFiles/News/CAERLS_Consultation_Paper_AppendixA.pdf, retrieved January 26, 2009.

5. Battles JB, Kaplan HS, Van der Schaaf TW, Shea CE. The attributes of medical event-reporting systems. Experience with a prototype medical event-reporting system for transfusion medicine. *Arch Pathol & Lab Med*, 1998;**122**(3):231–8. (See the *Eindhoven Classification Model for Medical Domain*, p.235.)

6. WHO World Alliance for Patient Safety. *WHO Draft Guidelines for Adverse Event Reporting and Learning Systems*. Geneva: World Health Organization (Document WHO/EIP/SPO/QPS/05.3), 2005.

7. Runciman WB. Shared meanings: preferred terms and definitions for safety and quality concepts. *MJA*, 2006;**184**(10):S41–S43.

8. United States Department of Veteran Affairs, National Center for Patient Safety. *Glossary of Patient Safety Terms*. 5 pages at http://www.va.gov/NCPS/glossary.html, retrieved November 9, 2008.

9. Agency for Healthcare Research and Quality (AHRQ), US Department of Health & Human Services. *Glossary*. 31 pages at http://www.webmm.ahrq.gov/popup_glossary.aspx, retrieved January 3, 2009, and 30 pages at http://www.wbmm.ahrq.gov/glossary.aspx, retrieved July 7, 2009.

10. Jenicek M. *Clinical Case Reporting in Evidence-based Medicine*. Second Edition. London and New York: Arnold and Oxford University Press, 2001.

11. Wilson W. *Causal Factor Tree Analysis*. 3 pages at http://www.bill-wilson.net/b56.html, retrieved June 19, 2009.

12. Jenicek M, Hitchcock DL. *Evidence-Based Practice. Logic and Critical Thinking in Medicine*. Chicago: American Medical Association (AMA Press), 2005.

13. *A Dictionary of Epidemiology*. Fifth Edition. Edited for the International Epidemiological Association by M. Porta. New York and Oxford: Oxford University Press, 2009.

14. Jenicek M. *Foundations of Evidence-Based Medicine*. Boca Raton, London, New York, Washington: The Parthenon Publishing Group Inc. (CRC Press), 2003.

15. Sasou K, Reason J. Team errors: definition and taxonomy. *Reliability Eng Syst Safety*, 1999;**65**(1):1–9.

16. Answers.com™. *Cognition*. 12 pages at http://www.answers.com/topic/cognition, retrieved November 12, 2008.

17. Bogner MS. Introduction. Pp. 1-12 in: *Human Error in Medicine*. Edited by MS Bogner. Hillsdale, NJ: Lawrence Erlbaum Associates, 1994.

18. WHO World Alliance for Patient Safety. *The Conceptual Framework for the International Classification for Patient Safety (ICPS). Version 1.0 for Use in Field Testing 2007–2008*. Geneva: World Health Organization, July 2007. 48 pdf pages at http://www.who.int/patientsafety/taxonomy/icps_download/en/print.html, retrieved February 4, 2009.

19. Scriven M, Paul R. Critical Thinking Community. A working definition of *critical thinking*. Available at http://lonestar.texas.net/~mseifert/crit2.html, retrieved June 2005.

20. Ennis RH. *Critical Thinking*. Upper Saddle River, NJ: Prentice Hall, 1996.

21. Institute of Medicine, Committee on Quality of Health Care in America. *To Err Is Human. Building a Safer Health System*. Edited by LT Kohn, JM Corrigan, and MS Donaldson. Washington, DC: National Academy Press, 2000.

22. Simpson K. *A Doctor's Guide to Court. A Handbook on Medical Evidence.* London: Butterworths, 1967.

23. Dowden B. *The Internet Encyclopedia of Philosophy. Fallacies.* 44 pages at http://www.iep.utm.edu/f/fallacy.htm, retrieved October 31, 2006.

24. Jenicek M. *Fallacy-Free Reasoning in Medicine. Improving Communication and Decision Making in Research and Practice.* Chicago: American Medical Association (AMA Press), 2009.

25. Wikipedia, the free encyclopedia. *Failure mode and effect analysis.* 7 pages at http://en.wikipedia.org/wiki/Failure_mode_and_effects_analysis, retrieved July 11, 2009.

26. Testability.com. *Reliability Terms.* 5 pages at http://www.testability.com/reference/Glossaries.aspx?Glossary=reliability, retrieved July 11, 2009.

27. Thiyagarajan Veluchamy Blog. *Glossary.* 29 pages at http:///thiyagarajan.wordpress.com/glossary/, retrieved July 11, 2009.

28. Wikipedia, the free encyclopedia. *Failure mode and effect analysis.* 4 pages at http://en.wikipedia.og/wiki/Failure_mode_and_effects_analysis, retrieved February 3, 2009.

29. American Society for Quality (ASQ). *Failure Modes and Effect Analysis (FMEA).* 3 pages at http://www.asq.org/learn-about-qyality/process-analysis-tools/overview/fmea.html, retrieved February 3, 2009.

30. Peters GA, Peters BJ. *Human Error: Causes and Control.* Boca Raton/London/New York: CRC/Taylor & Francis Group, 2006.

31. Runciman W, Hibbert P, Thomson R, Van der Schaaf T, Sherman H, Lewalle P. Towards an international classification of patient safety: key concepts and terms. *Int J Qual Health Care,* 2009;**21**(1):18–26.

32. WHO World Alliance for Patient Safety. *The Conceptual Framework for the International Classification for Patient Safety (ICPS). Version 1.0 for Use in Field Testing 2007-2008.* Geneva: World Health Organization, July 2007. 48 pdf pages at http://www.ismp-canada.org/definitions.htm, retrieved February 4, 2009.

33. Institute for Safe Medication Practices Canada; c2000–2006. *Definitions of Terms.* 3 pages at http://www.ismp-canada.org/definitions.htm, retrieved September 7, 2009.

34. Gold JA. The 5 Million Lives Campaign: Preventing medical harm in Wisconsin and the nation. *Wisconsin Med J,* 2008;**107**(5):270–1.

35. Sheridan-Leos N, Schulmeister L, Hartfant S. Failure mode and effect analysis™. A technique to prevent chemotherapy errors. *Clin J Oncol Nurs,* 2006;**10**(3):393–401.

36. Reason J. *Human Error.* Cambridge and New York: Cambridge University Press, 1990.

37. Steel K, Gertman PM, Crescenzi C, Anderson J. Iatrogenic illness on a general medical service at a university hospital. *N Engl J Med,* 1981;**304**:838–42.

38. Croskerry P, Cosby KS, Schenkel SM, Wears RL. *Patient Safety in Emergency Medicine.* Philadelphia, PA: Wolters Kluwer | Lippincott Williams & Wilkins, 2009.

39. Busse DK. *Cognitive Error Analysis in Accident and Incident Investigation in Safety-Critical Domains.* A PhD Thesis. Glasgow: University of Glasgow, Department of Computing Science, September 2002. 288 + XIII pages at www.dcs.gla.ac.uk/~johnosn/papers/Phd_DBusse.pdf, retrieved October 25, 2008.

40. Baker R, Grosso F, Heinz C, Sharpe, G Beardwood J, et al. Review of provincial, territorial and federal legislation and policy related to the reporting and review of adverse events in healthcare in Canada. Pp. 133–137 in: White JL. *Adverse Event Reporting and Learning Systems: A Review of the Relevant Literature.* Edmonton and Ottawa: Canadian Patient Safety Institute, June 25, 2007. 220 pdf pages at www.patientsafetyinstitute.ca/uploadedFiles/News/CAERLS_Consultation_Paper_AppendixA.pdf, retrieved January 26, 2009.

41. Yogis JA, Cotter C. *Barron's Canadian Law Dictionary.* 6th Edition. Hauppauge, NY: Barron's Educational Series, Inc., 2009.

42. Thomsen CJ. *The Scope of the Medication Error Problem.* 7 html pages at http://74.125.113.132/search?q=cache:iJ_sGynyyTWJ:www.thethomsengroup.com/TTGI..., retrieved January 28, 2009.

43. National Coordinating Council for Medication Error Reporting and Prevention. *What Is a Medication Error?* 1 page at http://www.nccmerp.org/aboutMedErrors.html?USP_Print=true&frame=lowerfrm, retrieved January 14, 2009.

44. Wu AW, Cavanaugh, TA, McPhee SJ et al. To tell the truth: ethical and practical issues in disclosing medical mistakes to patients. *J Gen Intern Med,* 1997;**12**:770–5.

45. College of Licensed Practical Nurses of Nova Scotia, College of Occupational Therapists of Nova Scotia, College of Physicians & Surgeons of Nova Scotia, College of Registered Nurses of Nova Scotia, Nova Scotia College of Pharmacists, Nova Scotia College of Physiotherapists. *Joint Position Statement on Patient Safety.* 8 pdf. pages at www.cpsns.ns.ca/2008-joint-patient-safety.pdf, retrieved January 27, 2009.

46. Strauss A, Corbin J. *Basics of Qualitative Research. Grounded Theory Procedures and Techniques.* Newbury Park: Sage Publications, 1990.

47. *New Technology and Human Error.* Edited by J Rasmussen, K Duncan, and J Leplat. Chichester and New York: John Wiley & Sons, 1987.

48. The Royal College of General Practitioners. *2. Significant event analysis.* 1 page at http://www.org.uk/practising_as_a_gp/distance_learning/egp2_update/learning_tool..., retrieved June 19, 2009.

49. *Human Error: Cause, Prediction, and Reduction.* Analysis and Synthesis by JW Senders and NP Moray. Hillsdale, NJ, Hove, London: Lawrence Erlbaum Associates, Publishers, 1991.

50. Cosby KS. Developing taxonomies for adverse events in emergency medicine. Chapter 5, pp. 58–69 in: Croskerry P, Cosby KS, Schenkel SM, Wears RL. *Patient Safety in Emergency Medicine.* Philadelphia/London/Sydney/Tokyo: Wolters Kluwer | Lippincott Williams & Wilkins, 2009.

51. Zhang J, Patel VL, Johnson TR, Shortliffe EH. A cognitive taxonomy of medical errors. *J Biomed Informatics,* 2004;**37**:193–204.

52. Donabedian A. Evaluating quality of medical care. *Milbank Mem Fund Quarterly,* 1966;**44**(Suppl.):166–206.

53. Wikipedia, the free encyclopedia. *System.* 6 pages at http://en.wikipedia.org/wiki/System, retrieved Jan 26, 2009.

54. Vincent C, Taylor-Adams S, Chapman JE, Hewett D, Prior S, Strange P, Tizzard A. How to investigate and analyze clinical incidents: Clinical Risk Unit and Association of Litigation and Risk Management Protocol. *BMJ,* 2000;**320**(18 March):777–81.

55. Wikipedia, the free encyclopedia. *Thought experiment. 12 pages at* http:// en.wikipedia.org/wiki/Thought_experiment, retrieved January 26, 2009.

56. Stanford Encyclopedia of Philosophy. *Thought Experiments.* March 25, 2007, revision. 14 pages at http://plato.dstanford.edu/entries/thought-experiment/, retrieved January 26, 2009.

57. Clapp JE. *Radom House Webster's Dictionary of Law.* New York: Random House, 2000.

58. Cacciabue PC. *Guide to Applying Human Factors Methods. Human Error and Accident Management in Safety Critical Systems.* London/Berlin/Heidelberg: Springer Verlag, 2004.

Appendix A:
List of Cognitive Biases

This Appendix is an author-reviewed version of the "List of Cognitive Biases" by *Wikipedia, the Free Encyclopedia* (retrieved August 29, 2009, from http://en.wikipedia.org/wiki/List_of_cognitive biases). Permission is granted under the terms of the GNU Free Documentation License, Version 1.2 or any later version published by the Free Software Foundation; with no invariant sections, with no front-cover texts, and with no back-cover texts. In terms of *Wikipedia Copyrights* (retrieved February 3, 2008, from http://en.wikipedia.org/wiki/Wikipedia:Copyrights), this Appendix is licensed under the GNU Free Documentation License. It uses the material from the Wikipedia article "List of Cognitive Biases."

In respect and conformity with Wikipedia rules and requirements, any copy of this Appendix can be made without restrictions.

Under the name of cognitive bias, this list contains several entries that are elsewhere in this book synonymous to some entities under the name of fallacies as defined and discussed in Appendix B, "List of Fallacies." With some overlaps, this list completes other lists quoted and referred to in this book. Synonymous or alternative terms and their definitions referred to, but not defined in those lists, were compiled based on information from other Wikipedia sites and kept in their original form.

Cognitive Biases: What They Are

A cognitive bias is a pattern of deviation in judgment that occurs in particular situations (see also cognitive distortion and the lists of thinking-related topics). Implicit in the concept of a "pattern of deviation" is a standard of comparison; this may be the judgment of people outside those particular situations, or may be a set of independently verifiable facts. The existence of some of these cognitive biases has been verified empirically in the field of psychology; others are widespread beliefs and may themselves be a consequence of cognitive bias.

Cognitive biases are instances of evolved mental behavior. Some are presumably adaptive, for example, because they lead to more effective actions or enable faster decisions. Others presumably result from a lack of appropriate mental mechanisms or from the misapplication of a mechanism that is adaptive under different circumstances. The next sections discuss the following four categories:

- Decision making and behavioral biases.
- Biases of probability and belief.
- Social biases.
- Memory errors (biases).

Decision Making and Behavioral Biases

Many of these biases are studied for how they affect belief formation, business decisions, and scientific research.

- **Bandwagon effect**: The tendency to do (or believe) things because many other people do (or believe) the same. Related to groupthink and herd behavior.
- **Base rate fallacy**: Ignoring available statistical data in favor of particulars.
- **Bias blind spot**: The tendency not to compensate for one's own cognitive biases.
- **Choice-supportive bias**: The tendency to remember one's choices as better than they actually were.
- **Confirmation bias**: The tendency to search for or interpret information in a way that confirms one's preconceptions.
- **Congruence bias**: The tendency to test hypotheses exclusively through direct testing, in contrast to tests of possible alternative hypotheses.
- **Contrast effect**: The enhancement or diminishing of a weight or other measurement when compared with a recently observed contrasting object.
- **Déformation professionnelle**: The tendency to look at things according to the conventions of one's own profession, forgetting any broader point of view.
- **Denomination effect:** The tendency to spend more money when it is denominated in small amounts (e.g., coins) rather than in large amounts (e.g., bills).
- **Distinction bias**: The tendency to view two options as more dissimilar when evaluating them simultaneously than when evaluating them separately.

- *Endowment effect*: "The fact that people often demand much more to give up an object than they would be willing to pay to acquire it."
- *Experimenter's* or **expectation bias**: The tendency for experimenters to believe, certify, and publish data that agree with their expectations for the outcome of an experiment, and to disbelieve, discard, or downgrade the corresponding weightings for data that appear to conflict with those expectations.
- *Extraordinarity bias*: The tendency to value an object more than others in the same category as a result of an extraordinarity of that object that does not, in itself, change the value.
- *Focusing effect*: Prediction bias occurring when people place too much importance on one aspect of an event; causes error in accurately predicting the utility of a future outcome.
- *Framing*: Using an approach or description of the situation or issue that is too narrow.
- *Framing effect*: Drawing different conclusions based on how data are presented.
- *Hyperbolic discounting*: The tendency for people to have a stronger preference for more immediate payoffs relative to later payoffs, where the tendency increases the closer to the present both payoffs are.
- *Illusion of control*: The tendency for human beings to believe they can control or at least influence outcomes that they clearly cannot.
- *Impact bias*: The tendency for people to overestimate the length or the intensity of the impact of future feeling states.
- *Information bias*: The tendency to seek information even when it cannot affect action.
- *Irrational escalation*: The tendency to make irrational decisions based on rational decisions in the past or to justify actions already taken.
- *Loss aversion*: "The disutility of giving up an object is greater than the utility associated with acquiring it" (see also *sunk cost effects* and *endowment effect*).
- *Mere exposure effect*: The tendency for people to express undue liking for things merely because they are familiar with them.
- *Moral credential effect*: The tendency of a track record of non-prejudice to increase subsequent prejudice.
- *Need for closure*: The need to reach a verdict in important matters; to have an answer and to escape the feeling of doubt and uncertainty. The personal context (time or social pressure) might increase this bias.
- *Neglect of probability*: The tendency to completely disregard probability when making a decision under uncertainty.
- *Not invented here*: The tendency to ignore that a product or solution already exists, because its source is seen as an "enemy" or as "inferior."

■ *Omission bias*: The tendency to judge harmful actions as worse, or less moral, than equally harmful omissions (inactions).

■ *Outcome bias*: The tendency to judge a decision by its eventual outcome instead of based on the quality of the decision at the time it was made.

■ *Planning fallacy*: The tendency to underestimate task completion times.

■ *Postpurchase rationalization:* The tendency to persuade oneself through rational argument that a purchase was a good value.

■ *Pseudo-certainty effect*: The tendency to make risk-averse choices if the expected outcome is positive but to make risk-seeking choices to avoid negative outcomes.

■ *Reactance*: The urge to do the opposite of what someone wants you to do out of a need to resist a perceived attempt to constrain your freedom of choice.

■ *Restraint bias*: The tendency to overestimate one's ability to show restraint in the face of temptation.

■ *Selective perception*: The tendency for expectations to affect perception.

■ *Semmelweis reflex*: The tendency to reject new evidence that contradicts an established paradigm.

■ *Status quo bias*: The tendency for people to like things to stay relatively the same (see also *loss aversion, endowment effect,* and *system justification*).

■ *Sunk cost effect ("loss aversion," sunk cost fallacy, sunk cost dilemma)*: A theory derived from economics and business decision-making stating that sunk costs (retrospective, past costs) that have already been incurred and cannot be recovered are relevant to an investment decision. For example, the cost of an unfinished research facility, a sunken cost, is used as a reason to finish the facility despite its questionable future projected benefit. Decisions depend on human and material resources already invested despite the final value of the activity (project) that is independent from the investment already made. In medicine, it may be a decision to pursue the treatment of a patient using a medication of questionable effectiveness because "the treatment of the patient using this medication already costs us a lot." (Summarized from other ad hoc Wikipedia Web sites.)

■ *Subject–expectancy effect*: A form of reactivity that occurs in scientific experiment or medical treatment when a research subject or patient expects a given result and therefore unconsciously affects the outcome, or reports the expected result. It can result in the subject experiencing the placebo effect or nocebo effect, depending on how the influence pans out.

■ *Von Restorff effect*: The tendency for an item that "stands out like a sore thumb" to be more likely to be remembered than other items.

■ *Wishful thinking*: The formation of beliefs and the making of decisions according to what is pleasing to imagine instead of by appeal to evidence or rationality.

■ *Zero-risk bias*: Preference for reducing a small risk to zero over a greater reduction in a larger risk.

Biases in Probability and Belief

Many of these biases are often studied for how they affect business and economic decisions and how they affect experimental research.

■ *Ambiguity effect*: The avoidance of options for which missing information makes the probability seem "unknown."

■ *Anchoring*: The tendency to rely too heavily, or "anchor," on a past reference or on one trait or piece of information when making decisions.

■ *Attentional bias*: Neglect of relevant data when making judgments of a correlation or association.

■ *Authority bias*: The tendency to value an ambiguous stimulus (e.g., an art performance) according to the opinion of someone who is seen as an authority on the topic.

■ *Availability heuristic*: Estimating what is more likely by what is more available in memory, which is biased toward vivid, unusual, or emotionally charged examples.

■ *Availability cascade*: A self-reinforcing process in which a collective belief gains more and more plausibility through its increasing repetition in public discourse (or "repeat something long enough and it will become true").

■ *Belief bias*: An effect where someone's evaluation of the logical strength of an argument is biased by the believability of the conclusion.

■ *Clustering illusion*: The tendency to see patterns where actually none exist.

■ *Capability bias*: The tendency to believe that the closer average performance is to a target, the tighter the distribution of the data set.

■ *Conjunction fallacy*: The tendency to assume that specific conditions are more probable than general ones.

■ *Gambler's fallacy*: The tendency to think that future probabilities are altered by past events, when in reality they are unchanged. Results from an erroneous conceptualization of the normal distribution. For example, "I've flipped heads with this coin five times consecutively, so the chance of tails coming out on the sixth flip is much greater than heads."

■ *Hawthorne effect*: The tendency of people to perform or perceive differently when they know that they are being observed.

■ *Hindsight bias*: Sometimes called the "I-knew-it-all-along" effect, the inclination to see past events as being predictable.

■ *Illusory correlation*: Beliefs that inaccurately suppose a relationship between a certain type of action and an effect.

■ *Ludic fallacy*: The analysis of chance related problems according to the belief that the unstructured randomness found in life resembles the structured randomness found in games, ignoring the non-Gaussian distribution of many real-world results.

■ *Neglect of prior base rates effect*: The tendency to neglect known odds when reevaluating odds in light of weak evidence.

■ *Observer-expectancy effect*: When a researcher expects a given result and therefore unconsciously manipulates an experiment or misinterprets data to find it (see also *subject-expectancy effect*).

■ *Optimism bias*: The systematic tendency to be overly optimistic about the outcome of planned actions.

■ *Ostrich effect*: Ignoring an obvious (negative) situation.

■ *Overconfidence effect*: Excessive confidence in one's own answers to questions. For example, for certain types of question, answers that people rate as "99% certain" turn out to be wrong 40% of the time.

■ *Positive outcome bias*: A tendency in prediction to overestimate the probability of good things happening to them (see also *wishful thinking, optimism bias,* and *valence effect*).

■ *Pareidolia*: Vague and random stimulus (often an image or sound) are perceived as significant (e.g., seeing images of animals or faces in clouds, the man in the moon, and hearing hidden messages on records played in reverse).

■ *Peak-end rule*: We judge our past experiences almost entirely on how they were at their *peak* (pleasant or unpleasant) and how they ended. Other information and the sum of an experience are not lost, but they are not used.

■ *Primacy effect*: The tendency to weigh initial events more than subsequent events.

■ *Recency effect*: The tendency to weigh recent events more than earlier events (see also *peak-end rule*).

■ Disregard of *regression toward the mean*: The tendency to expect extreme performance to continue.

■ *Reminiscence bump*: The effect that people tend to recall more personal events from adolescence and early adulthood than from other lifetime periods.

■ *Rosy retrospection*: The tendency to rate past events more positively than they had actually rated them when the event occurred.

■ *Selection bias*: A distortion of evidence or data that arises from the way that the data are collected.

■ *Stereotyping*: Expecting a member of a group to have certain characteristics without having actual information about that individual.

- *Subadditivity effect*: The tendency to judge probability of the whole to be less than the probabilities of the parts.
- *Subject–expectancy effect*: A form of reactivity that occurs in scientific experiment or medical treatment when a research subject or patient expects a given result and therefore unconsciously affects the outcome, or reports the expected result. It can result in the subject experiencing the placebo effect or nocebo effect, depending on how the influence pans out.
- *Subjective validation*: Perception that something is true if a subject's belief demands it to be true. Also assigns perceived connections between coincidences.
- *Telescoping effect*: The effect that recent events appear to have occurred more remotely and remote events appear to have occurred more recently.
- *Texas sharpshooter fallacy*: The fallacy of selecting or adjusting a hypothesis after the data are collected, making it impossible to test the hypothesis fairly. Refers to the concept of firing shots at a barn door, drawing a circle around the best group, and declaring that to be the target.
- *Valence effect (wishful thinking, self-serving bias)*: A tendency for people to simply overestimate the likelihood of good things happening rather than bad things. *Valence* refers to the *positive* or *negative* emotional charge something has and also to over-predicting the likelihood of positive effects happening to *ourselves* relative to others.

Social (most of them *attribution*) Biases

- *Actor–observer bias:* The tendency for explanations of other individuals' behaviors to overemphasize the influence of their personality and underemphasize the influence of their situation. However, this is coupled with the opposite tendency for the self in that explanations for our own behaviors overemphasize the influence of our situation and underemphasize the influence of our own personality (see also *fundamental attribution error*).
- *Dunning-Kruger effect*: "...When people are incompetent in the strategies they adopt to achieve success and satisfaction, they suffer a dual burden: Not only do they reach erroneous conclusions and make unfortunate choices, but their incompetence robs them of the ability to realize it. Instead, they are left with the mistaken impression that they are doing just fine" (see also *Lake Wobegon effect* and *overconfidence effect*).
- *Egocentric bias*: Occurs when people claim more responsibility for themselves for the results of a joint action than an outside observer would.
- *Forer effect* (or *Barnum effect*): The tendency to give high accuracy ratings to descriptions of their personality that supposedly are tailored spe-

cifically for them but are in fact vague and general enough to apply to a wide range of people (e.g., horoscopes).

- **False consensus effect:** The tendency for people to overestimate the degree to which others agree with them.

- **Fundamental attribution error:** The tendency for people to overemphasize personality-based explanations for behaviors observed in others while underemphasizing the role and power of situational influences on the same behavior (see also *actor-observer bias, group attribution error, positivity effect,* and *negativity effect*).

- **Group-serving bias:** Identical to *self-serving bias (vide infra)* except that it takes place between groups rather than individuals, under which group members make dispositional attributions for their group successes and situational attributions for group failures, and vice versa for outsider groups.

- **Halo effect:** The tendency for people's positive or negative traits to "spill over" from one area of their personality to another in others' perceptions of them (see also *physical attractiveness stereotype*).

- **Herd instinct:** Common tendency to adopt the opinions and follow the behaviors of the majority to feel safer and to avoid conflict.

- **Illusion of asymmetric insight:** People perceive their knowledge of their peers to surpass their peers' knowledge of them.

- **Illusion of transparency:** People overestimate others' ability to know them, and they also overestimate their ability to know others.

- **Illusory superiority:** Perceiving oneself as having desirable qualities to a greater degree than other people. Also known as superiority bias (see also *Lake Wobegon effect*).

- **Ingroup bias:** The tendency for people to give preferential treatment to others they perceive to be members of their own groups.

- **Just-world phenomenon:** The tendency for people to believe that the world is "just" and therefore people "get what they deserve."

- **Lake Wobegon effect:** The phenomenon that a supermajority of people report themselves as above average in desirable qualities (see also *worse-than-average effect, illusory superiority,* and *optimism bias*).

- **Money illusion:** An irrational notion that the arbitrary values of currency, fiat or otherwise, have an actual immutable value.

- **Notational bias:** A form of cultural bias in which a notation induces the appearance of a nonexistent natural law.

- **Outgroup homogeneity bias:** Individuals see members of their own group as being relatively more varied than members of other groups.

- **Physical attractiveness stereotype:** The tendency to assume that people who are physically attractive also possess other socially desirable personality traits.

- **Projection bias:** The tendency to unconsciously assume that others share the same or similar thoughts, beliefs, values, or positions.
- **Self-serving bias**: The tendency to claim more responsibility for successes than failures. It may also manifest itself as a tendency for people to evaluate ambiguous information in a way beneficial to their interests (see also *group-serving bias*).
- **Self-fulfilling prophecy:** The tendency to engage in behaviors that elicit results that will (consciously or not) confirm our beliefs.
- **System justification**: The tendency to defend and bolster the status quo. Existing social, economic, and political arrangements tend to be preferred and alternatives disparaged sometimes, even at the expense of individual and collective self-interest (see also *status quo bias*).
- **Trait ascription bias:** The tendency for people to view themselves as relatively variable in terms of personality, behavior, and mood while viewing others as much more predictable.
- **Ultimate attribution error**: Similar to the fundamental attribution error, in this error a person is likely to make an internal attribution to an entire group instead of the individuals within the group.
- **Worse-than-average effect**: The human tendency to underestimate one's achievements and capabilities in relation to others.

Memory Errors (Biases)

- **Consistency bias**: Incorrectly remembering one's past attitudes and behavior as resembling present attitudes and behavior.
- **Cryptomnesia**: A form of *misattribution* where a memory is mistaken for imagination.
- **Egocentric bias:** Recalling the past in a self-serving manner (e.g., remembering one's exam grades as being better than they were; remembering a caught fish as being bigger than it was).
- **False memory:** Confusion of imagination with memory or the confusion of true memories with false memories.
- **Hindsight bias:** Filtering memory of past events through present knowledge so that those events look more predictable than they actually were; also known as the "I-knew-it-all-along effect."
- **Self-serving bias:** Perceiving oneself responsible for desirable outcomes but not responsible for undesirable ones.
- **Suggestibility:** A form of *misattribution* where ideas suggested by a questioner are mistaken for memory.

References

Baron J. *Thinking and Deciding* (3rd ed.), New York: Cambridge University Press, 2000.

Bishop MA, Trout D. *Epistemology and the Psychology of Human Judgment*, New York: Oxford University Press, 2004.

Gilovich T. *How We Know What Isn't So: The Fallibility of Human Reason in Everyday Life*, New York: Free Press, 1993.

Gilovich T, Griffin D, Kahneman D. *Heuristics and Biases: The Psychology of Intuitive Judgment*, Cambridge, UK: Cambridge University Press, 2002.

Greenwald AG. The totalitarian ego: Fabrication and revision of personal history. *Am Psychol,* 1980; **35**(7):603–18.

Kahneman D, Slovic P, Tversky A. *Judgment under Uncertainty: Heuristics and Biases*, Cambridge, UK: Cambridge University Press, 1982.

Kahneman D, Knetsch JL, Thaler RH. Anomalies: The endowment effect, loss aversion, and status quo bias. *J Econ Persp*, 1991; **5**(1):193–206.

Plous S. *The Psychology of Judgment and Decision Making*, New York: McGraw-Hill, 1993.

Schacter, DL. The seven sins of memory: Insights from psychology and cognitive neuroscience. *Am Psychol*, 1999; **54**(3):182–203.

Tetlock PE, Philip E. *Expert Political Judgment: How Good Is It? How Can We Know?* Princeton, NJ: Princeton University Press, 2005.

Virine L, Trumper M. *Project Decisions: The Art and Science*, Vienna, VA: Management Concepts, 2007.

Appendix B:
List of Fallacies

This Appendix is an author-reviewed version of the "List of Fallacies" by *Wikipedia, the Free Encyclopedia* (retrieved August 29, 2009, from http://en.wikipedia.org/wiki/List_of_fallacies). Permission is granted under the terms of the GNU Free Documentation License, Version 1.2 or any later version published by the Free Software Foundation; with no invariant sections, with no front-cover texts, and with no back-cover texts. In terms of *Wikipedia Copyrights* (retrieved February 3, 2008, from http://en.wikipedia.org/wiki/Wikipedia:Copyrights), this Appendix is licensed under the GNU Free Documentation License. It uses the material from the Wikipedia article "List of Cognitive Biases."

In respect and conformity with *Wikipedia* rules and requirements, any copy of this Appendix can be made without restrictions.

Under the name of cognitive bias, this list contains several entries that are elsewhere in this book synonymous to some entities under the name of cognitive bias as defined and commented in "Appendix A: List of Cognitive Biases." With some overlaps, this list completes the other lists quoted and referred to in this book, including the following categories:

- Formal fallacies.
- Propositional fallacies.
- Quantificational fallacies.
- Formal syllogistic fallacies.
- Informal fallacies.
- Faulty generalizations.
- Red herring fallacies.
- Conditional and questionable fallacies.

Formal Fallacies

Formal fallacies are arguments that are fallacious due to an error in their form or technical structure.[1] All formal fallacies are specific types of non sequiturs:

- **Ad hominem**: an argument that attacks the person who holds a view or advances an argument, rather than commenting on the view or responding to the argument.
- **Appeal to probability**: assumes that because something *could* happen, it is inevitable that it *will* happen. This is the premise on which Murphy's Law is based.
- **Argument from fallacy**: if an argument for some conclusion is fallacious, then the conclusion must be false.
- **Bare assertion fallacy**: premise in an argument is assumed to be true purely because it says that it is true.
- **Base rate fallacy**: using weak evidence to make a probability judgment without taking into account known empirical statistics about the probability.
- **Conjunction fallacy**: assumption that an outcome simultaneously satisfying multiple conditions is more probable than an outcome satisfying a single one of them.
- **Correlative based fallacies**
  - **Denying the correlative**: where attempts are made at introducing alternatives where there are none.
  - **Suppressed correlative**: where a correlative is redefined so that one alternative is made impossible.
- **Fallacy of necessity**: a degree of unwarranted necessity is placed in the conclusion based on the necessity of one or more of its premises.
- **False dilemma (false dichotomy)**: where two alternative statements are held to be the only possible options, when in reality there are more.
- **If-by-whiskey**: An answer that takes side of the questioner's suggestive question.
- **Ignoratio elenchi** (irrelevant conclusion or irrelevant thesis, ignorance of refutation): Presenting an argument that may in itself be valid, but does not address the issue in question.
- **Homunculus fallacy**: where a "middle-man" is used for explanation, this usually leads to regressive middle-man. Explanations without actually explaining the real nature of a function or a process.
- **Masked man fallacy**: the substitution of identical designators in a true statement can lead to a false one.
- **Naturalistic fallacy**: a fallacy that claims that if something is natural, then it is "good" or "right."

- *Nirvana fallacy*: when solutions to problems are said not to be right because they are not perfect.
- *Negative proof fallacy*: that, because a premise cannot be proven false, the premise must be true; or that, because a premise cannot be proven true, the premise must be false.
- *Package-deal fallacy*: consists of assuming that things often grouped together by tradition or culture must always be grouped that way.
- *Red Herring*: also called a "*fallacy of relevance*." This occurs when the speaker is trying to distract the audience by arguing some new topic, or just generally going off topic with an argument (see also as separate category).

Propositional Fallacies

- *Affirming a disjunct*: concluding that one logical disjunction must be false because the other disjunct is true; *A or B; A; therefore not B.*
- *Affirming the consequent*: the antecedent in an indicative conditional is claimed to be true because the consequent is true; *if A, then B; B, therefore A.*
- *Denying the antecedent*: the consequent in an indicative conditional is claimed to be false because the antecedent is false; *if A, then B; not A, therefore not B.*

Quantificational Fallacies

- *Existential fallacy*: an argument has two universal premises and a particular conclusion, but the premises do not establish the truth of the conclusion.
- *Proof by example*: where things are proven by giving an example.

Formal Syllogistic Fallacies (structural flaws)

- *Affirmative conclusion from a negative premise*: when a categorical syllogism has a positive conclusion, but at least one negative premise.
- *Fallacy of exclusive premises*: a categorical syllogism that is invalid because both of its premises are negative.
- *Fallacy of four terms*: a categorical syllogism has four terms.
- *Illicit major*: a categorical syllogism that is invalid because its major term is undistributed in the major premise but distributed in the conclusion.
- *Fallacy of the undistributed middle*: the middle term in a categorical syllogism is not distributed.

Informal Fallacies

Informal fallacies are arguments that are fallacious for reasons other than structural ("formal") flaws.

- *Argument from repetition (argumentum ad nauseam)*: signifies that it has been discussed extensively (possibly by different people) until nobody cares to discuss it anymore.
- *Appeal to ridicule*: a specific type of appeal to emotion where an argument is made by presenting the opponent's argument in a way that makes it appear ridiculous.
- *Argument from ignorance ("appeal to ignorance")*: The fallacy of assuming that something is true/false because it has not been proven false/true. For example: "The student has failed to prove that he didn't cheat on the test, therefore he must have cheated on the test."
- *Begging the question ("petitio principii")*: where the conclusion of an argument is implicitly or explicitly assumed in one of the premises.
- *Burden of proof*: refers to the extent to which, or the level of rigor with which, it is necessary to establish, demonstrate, or prove something for it to be accepted as true or reasonable to believe.
- *Circular cause and consequence*: where the consequence of the phenomenon is claimed to be its root cause.
- *Continuum fallacy (fallacy of the beard)*: appears to demonstrate that two states or conditions cannot be considered distinct (or do not exist at all) because between them there exists a continuum of states. According to the fallacy, differences in quality cannot result from differences in quantity.
- *Correlation does not imply causation (cum hoc ergo propter hoc)*: a phrase used in the sciences and the statistics to emphasize that correlation between two variables does not imply that one causes the other.
- *Equivocation (No true Scotsman)*: the misleading use of a term with more than one meaning (by glossing over which meaning is intended at a particular time).
- *Fallacies of distribution*
  - *Division*: where one reasons logically that something true of a thing must also be true of all or some of its parts.
  - *Ecological fallacy*: inferences about the nature of specific individuals are based solely upon aggregate statistics collected for the group to which those individuals belong.
- *Fallacy of many questions (complex question, fallacy of presupposition, loaded question, plurium interrogationum)*: someone asks a question that presupposes something that has not been proven or accepted by

all the people involved. This fallacy is often used rhetorically, so that the question limits direct replies to those that serve the questioner's agenda.

■ *Fallacy of the single cause ("joint effect," or "causal oversimplification")*: occurs when it is assumed that there is one, simple cause of an outcome when in reality it may have been caused by a number of only jointly sufficient causes.

■ *False attribution*: occurs when an advocate appeals to an irrelevant, unqualified, unidentified, biased or fabricated source in support of an argument.
  - *Contextomy (quoting out of context)*: refers to the selective excerpting of words from their original linguistic context in a way that distorts the source's intended meaning.

■ *False compromise/middle ground*: asserts that a compromise between two positions is correct.

■ *Gambler's fallacy*: the incorrect belief that the likelihood of a random event can be affected by or predicted from other, independent events.

■ *Historian's fallacy*: occurs when one assumes that decision makers of the past viewed events from the same perspective and having the same information as those subsequently analyzing the decision. It is not to be confused with *presentism*, a mode of historical analysis in which present-day ideas (such as moral standards) are projected into the past.

■ *Incomplete comparison*: where not enough information is provided to make a complete comparison.

■ *Inconsistent comparison*: where different methods of comparison are used, leaving one with a false impression of the whole comparison.

■ *Intentional fallacy*: addresses the assumption that the meaning intended by the author of a literary work is of primary importance.

■ *Loki's wager*: the unreasonable insistence that a concept cannot be defined, and therefore cannot be discussed.

■ *Moving the goalpost (raising the bar)*: argument in which evidence presented in response to a specific claim is dismissed and some other (often greater) evidence is demanded.

■ *Perfect solution fallacy*: where an argument assumes that a perfect solution exists and/or that a solution should be rejected because some part of the problem would still exist after it was implemented.

■ *Post hoc ergo propter hoc*: also known as false cause, coincidental correlation or correlation not causation.

■ *Proof by verbosity (proof by intimidation, argumentum verbosium)*: Persuasion by overwhelming the reader or listener with such a volume of material that a proposition sounds plausible, superficially appears to be well researched, or is so laborious to untangle and check that the proposition is allowed to slide by unchallenged. (WC Field's statement: *If you can't dazzle them with brilliance, baffle them with bull.*)

- **Prosecutor's fallacy**: a low probability of false matches does not mean a low probability of *some* false match being found.
- **Psychologist's fallacy**: occurs when an observer presupposes the objectivity of his own perspective when analyzing a behavioral event.
- **Regression fallacy**: ascribes cause where none exists. The flaw is failing to account for natural fluctuations. It is frequently a special kind of the post hoc fallacy.
- **Reification (hypostatization)**: a fallacy of ambiguity, when an abstraction (abstract belief or hypothetical construct) is treated as if it were a concrete, real event or physical entity. In other words, it is the error of treating as a "real thing" something which is not a real thing, but merely an idea.
- **Retrospective determinism**: it happened so it was bound to.
- **Special pleading**: where a proponent of a position attempts to cite something as an exemption to a generally accepted rule or principle without justifying the exemption.
- **Suppressed correlative**: an argument that tries to redefine a correlative (two mutually exclusive options) so that one alternative encompasses the other, thus making one alternative impossible.
- **Wrong direction**: where cause and effect are reversed. The cause is said to be the effect and vice versa.

Faulty Generalizations

- **Accident (fallacy)**: when an exception to the generalization is ignored.
- **Cherry picking**: act of pointing at individual cases or data that seem to confirm a particular position, while ignoring a significant portion of related cases or data that may contradict that position.
- **Composition**: where one infers that something is true of the whole from the fact that it is true of some (or even every) part of the whole.
- **Dicto simpliciter** (spoken simply, from a universal rule, sweeping generalization): Making an "universal" statement and expecting to be true of every specific case. An acceptable exception is ignored or eliminated.
  - **Accident (a dicto simpliciter ad dictum secundum quid)**: when an acceptable exception is ignored or eliminated.
  - **Converse accident (a dicto secundum quid ad dictum simpliciter)**: when an exception to a generalization is wrongly called for.

- **False analogy**: false analogy consists of an error in the substance of an argument (the content of the analogy itself), not an error in the logical structure of the argument.
- **Hasty generalization (fallacy of insufficient statistics, fallacy of insufficient sample, fallacy of the lonely fact, leaping to a conclusion, hasty induction, secundum quid)**: Faulty generalization by reaching an inductive generalization based on insufficient evidence.
- **Loki's wager**: insistence that because a concept cannot be clearly defined, it cannot be discussed.
- **Misleading vividness**: involves describing an occurrence in vivid detail, even if it is an exceptional occurrence, to convince someone that it is a problem.
- **Overwhelming exception (hasty generalization)**: It is a generalization which is accurate, but comes with one or more qualifications which eliminate so many cases that what remains is much less impressive than the initial statement might have led one to assume.
- **Pathetic fallacy**: when an inanimate object is declared to have characteristics of animate objects.
- **Spotlight fallacy**: when a person uncritically assumes that all members or cases of a certain class or type are like those that receive the most attention or coverage in the media.
- **Thought-terminating cliché**: a commonly used phrase, sometimes passing as folk wisdom, used to quell cognitive dissonance.

Red Herring Fallacies

A red herring is an argument, given in response to another argument, which does not address the original issue.

- **Ad hominem**: attacking the personal instead of the argument. A form of this is **reductio ad Hitlerum**.
- **Argumentum ad baculum ("appeal to force," "appeal to the stick")**: where an argument is made through coercion or threats of force towards an opposing party.
- **Argumentum ad populum ("appeal to belief," "appeal to the majority," "appeal to the people")**: where a proposition is claimed to be true solely because many people believe it to be true.
- **Association fallacy (guilt by association)**: Asserting that qualities of one thing are inherently qualities of another, merely by irrelevant association. (A special case of red herring fallacy that can be based on an appeal to emotion; see below.)

- **Appeal to authority**: where an assertion is deemed true because of the position or authority of the person asserting it.
- **Appeal to consequences**: a specific type of appeal to emotion where an argument that concludes a premise is either true or false based on whether the premise leads to desirable or undesirable consequences for a particular party.
- **Appeal to emotion**: where an argument is made due to the manipulation of emotions, rather than the use of valid reasoning.
  - **Appeal to fear**: a specific type of appeal to emotion where an argument is made by increasing fear and prejudice towards the opposing side.
  - **Wishful thinking**: a specific type of appeal to emotion where a decision is made according to what might be pleasing to imagine, rather than according to evidence or reason.
  - **Appeal to spite**: a specific type of appeal to emotion where an argument is made through exploiting people's bitterness or spite toward an opposing party.
  - **Appeal to flattery**: a specific type of appeal to emotion where an argument is made due to the use of flattery to gather support.
- **Appeal to motive**: where a premise is dismissed by calling into question the motives of its proposer.
- **Appeal to novelty**: where a proposal is claimed to be superior or better solely because it is new or modern.
- **Appeal to poverty (argumentum ad lazarum)**: thinking a conclusion is correct because the speaker is financially poor or incorrect because the speaker is financially wealthy.
- **Appeal to wealth (argumentum ad crumenam)**: concluding that a statement is correct because the speaker is rich or that a statement is incorrect because the speaker is poor.
- **Argument from silence (argumentum ex silentio)**: a conclusion based on silence or lack of contrary evidence.
- **Appeal to tradition**: where a thesis is deemed correct on the basis that it has a long-standing tradition behind it.
- **Chronological snobbery**: where a thesis is deemed incorrect because it was commonly held when something else, clearly false, was also commonly held.
- **Genetic fallacy**: where a conclusion is suggested based solely on something or someone's origin rather than its current meaning or context. This overlooks any difference to be found in the present situation, typically transferring the positive or negative esteem from the earlier context.
- **Judgmental language**: insultive or pejorative language to influence the recipient's judgment.

- ***Poisoning the well***: where adverse information about a target is pre-emptively presented to an audience, with the intention of discrediting or ridiculing everything that the target person is about to say.
- ***Sentimental fallacy***: it would be more pleasant if; therefore it ought to be; therefore it is.
- ***Straw man argument***: based on misrepresentation of an opponent's position.
- ***Style over substance fallacy***: occurs when one emphasizes the way in which the argument is presented, while marginalizing (or outright ignoring) the content of the argument.
- ***Texas sharpshooter fallacy***: information that has no relationship is interpreted or manipulated until it appears to have meaning.
- ***Two wrongs make a right***: occurs when it is assumed that if one wrong is committed, another wrong will cancel it out.
- ***Tu quoque***: the argument states that a certain position is false or wrong or should be disregarded because its proponent fails to act consistently in accordance with that position.

Conditional or Questionable Fallacies

- ***Definist fallacy***: involves the confusion between two notions by defining one in terms of the other.
- ***Luddite fallacy***: related to the belief that labor-saving technologies increase unemployment by reducing demand for labor.
- ***Broken window fallacy***: an argument that disregards hidden costs associated with destroying property of others.
- ***Slippery slope***: argument states that a relatively small first step inevitably leads to a chain of related events culminating in some significant impact.

Reference

Gupta B. *Perceiving in Advaita Vedanta: Epistemological Analysis and Interpretation.* New Delhi: Motilal Banrsidass, 1995 (p. 197).

More about medicine-related fallacies may be found in:

Jenicek M. *Fallacy-Free Reasoning in Medicine. Improving Communication and Decision Making in Research and Practice.* Chicago: American Medical Association Press, 2009.

Appendix C: Medical Error and Harm–Related Case Report*

B. A. Liang, M.D., Ph.D., J.D.

Professor and Director
Institute of Health Law Studies
California Western School of Law
University of California San Diego School of Medicine

The Case

A young woman, about 30 years of age, was injured in an automobile collision. She was brought to the emergency department (ED) via ambulance, where she was found to be suffering internal bleeding with life-threatening blood loss. After examination, physicians advised her that without transfusion of one to two units of blood within a very short time, she would die. The patient refused the transfusion, stating that her religion forbids it and that she understood the consequences. The ED staff deemed her to be competent and was ready to comply with her wishes.

---

* "No Blood, Please," *Clinical Ethics,* May 2004. AHRQ WebM&M [serial online]. Available at http://www.webmm.ahrq.goc/case.aspx?caseID=59 (with permission).

At about the same time that she was undergoing examination, both her parents and her minister arrived. The parents asserted that their daughter had recently converted to her new faith only weeks before and therefore did not fully understand why the religion forbids blood transfusion nor the consequences of her decision. The minister, on the other hand, stated that the woman converted to this religion with the full knowledge of its tenets and was well aware of the consequences of her decision. He stated that, at the time of her conversion, she swore an oath that she would live by the tenets of the faith; that oath contained language forbidding blood transfusions. As these discussions unfolded, the woman lost consciousness. The ED staff reversed their previous decision, and transfused two units of blood into the unconscious woman. She was then taken to surgery.

The woman recovered from her injuries. She and the minister of her church sued the hospital generally and ED staff specifically. The judgment ruled in her favor, saying that the hospital and ED staff violated her civil rights and interfered with her ability to make her own decisions.

The Commentary

This case illustrates how the law can interface with patient safety efforts, and highlights the need to pay close attention to patients' wishes, regardless of the potential clinical outcome.

As a primary matter, all patients have the constitutional right to determine what shall and shall not be done to them.[1] This right extends to any treatment that may save the patient's life. This includes blood transfusions, which are particularly important in circumstances involving Jehovah's Witness patients.[1,2] Only in emergency circumstances where a patient lacks capacity to consent can a provider transfuse blood without patient consent; courts have allowed providers to perform such treatment on the basis of assumed consent and public policy rationales.[1,3,4] In circumstances involving children, courts have both allowed and disallowed blood transfusion treatment recommendations when parents object on religious grounds.[1]

In this case, the patient was in fact cogent and of majority age; hence, her providers must follow her directives regarding her care. Unfortunately, although providers here were acting in good faith, any patient's care decisions are constitutionally protected. Therefore, transfusion of blood to her was in direct violation of her legal rights. In this situation, the hospital, emergency room, and individual providers can be liable for such actions, including actions for informed consent violation and civil battery. This latter cause of action is extremely important, since battery claims are amenable to punitive damages.[1] In the eyes of the law, it is irrelevant that the care provided did in fact save the patient's life.

Patients' choices of allowing and refusing treatment may present significant safety challenges. If patient preferences conflict with provider knowledge of appropriate care, these choices create additional sources of failure. Indeed, in this case, the decision to transfuse was a medical error. Many providers, such as emergency department, laboratory medicine, and anesthesia staff, will likely encounter such circumstances. Clear policies and procedures along with communication regarding protocol requirements in these situations should be extant and part of continuous safety training. It is not clear whether these components existed, were available, or were known to providers involved in this case. As a safety matter, root causes should be assessed to create system improvement to reduce the likelihood or mitigate potential recurrence of such an error. As a legal matter, to avoid such lawsuits, protocols should expressly indicate that, when faced with conflict between patient and provider preferences for care, the specific objections of the patient, recommendations of the provider, and competency assessments should be documented clearly by the provider. If possible, it would be prudent to have the patient indicate in his or her own words that he or she objects to the proposed treatment and is refusing consent; the patient should then sign that document. Moreover, at least one witness other than the direct care provider should be present and sign the patient's statement. In the event that no policy or procedure is extant, providers should contact their general counsel to determine what action(s) to take.

Hence, in this case, providers should not have transfused the patient on both ethical and legal grounds. Policies, procedures, and provider education should be implemented so that occurrence of similar medical errors is minimized or potentially mitigated by medical staff. This approach is consistent with patient-centered care and respect for patient autonomy as well as reduces the risk of litigation and associated damages that could be associated with such actions.

Take-Home Points

- Patients have the constitutional right to accept or refuse care.
- In the eyes of the law, whether the care provided saved the patient's life is irrelevant.
- If patient preferences and provider recommendations conflict, the objections, recommendations, and competency assessment should be explicitly and clearly documented.
- Since patient preferences sometimes create safety challenges, clear policies and procedures should be in place regarding such situations, and related information should be included in patient safety education efforts.

References

1. Liang BA. Informed consent. In: Liang BA. *Health Law & Policy*. Boston, MA: Butterworth-Heinemann; 2000.
2. *In the Matter of Melideo*, 390 N.Y.S.2d 253 (N.Y.Sup.Ct. 1976).
3. *John F. Kennedy Mem. Hosp. v. Heston*, 279 A.2d 670 (N.J. 1971).
4. Hartman KM, Liang BA. Exceptions to informed consent in emergency medicine. *Hosp Physician*,1999;**35**:53–9.

About the Author

Milos Jenicek is a Canadian physician. He is currently professor (part-time) in the Department of Clinical Epidemiology and Biostatistics of the Michael G DeGroote School of Medicine, Faculty of Health Sciences, McMaster University, Hamilton, Ontario, Canada. He is also professor emeritus at the Université de Montréal and is adjunct professor at the McGill University Faculty of Medicine in Montreal, Quebec, Canada. In 2009, he was elected Fellow of The Royal Society of Medicine, London.

Dr. Jenicek was awarded his M.D. in 1959 and his Ph.D. from Charles University in Prague in 1965. Later, he received postgraduate clinical training at McGill University Teaching Hospitals. He is a Licentiate of the Medical Council of Canada (LMCC), a Fellow of the Royal College of Physicians and Surgeons of Canada (FRCPC), and a specialist of the Province of Quebec (CSPQ, community medicine). He contributes to the evolution of epidemiology as a general method of objective reasoning and decision making in medicine.

He also authored the first textbook on meta-analysis in medicine recognized by the Oxford James Lind Library, preceded by several books written in French on modern epidemiology. Until 1991, he was member of the Board of Examiners of the Medical Council of Canada (Committee on Preventive Medicine). Dr. Jenicek is also a consultant for various national and international public and private bodies as well as an editorial consultant for several medical journals.

In addition to numerous scientific papers, the author has published 12 textbooks. *Introduction to Epidemiology* (in French, 1975); *Epidemiology: Principles, Techniques, Applications* (in French with R. Cléroux, 1982, and in Spanish, 1987); *Clinical Epidemiology: Clinimetrics* (in French with R. Cléroux, 1985); and *Meta-Analysis in Medicine: Evaluation and Synthesis of Clinical and Epidemiological Information* (in French, 1987), the first textbook on meta-analysis in medicine. *Epidemiology: The Logic of Modern Medicine* (EPIMED International, 1995) was also published in Spanish (1996) and Japanese (1998). His sixth book, *Medical Casuistics: Proper Reporting of Clinical Cases* (in French, 1997) is again produced jointly by Canadian (EDISEM) and French (Maloine) publishers. *Clinical Case Reporting in Evidence-Based Medicine* (Butterworth Heinemann, 1999) has been printed with an expanded second edition in English (Arnold, 2001), Italian (2001), Korean (2002), and Japanese (2002). *Foundations of Evidence-Based Medicine* was published in 2003 by Parthenon Publishing/CRC Press. The American Medical Association Press published *Evidence-Based Practice: Logic and Critical Thinking in Medicine* (with DL Hitchcock, 2005); *A Physician's Self-Paced Guide to Critical Thinking* (2006); and *Fallacy-Free Reasoning in Medicine: Improving Communication and Decision Making in Research and Practice* (2009).

Dr. Jenicek's current interests include development of methodology and applications of logic and critical thinking in health sciences, enhancement of evidence-based medicine and evidence-based public health, health policies and program evaluation, and decision oriented (bedside) clinical research.

Index

Printed and bound by CPI Group (UK) Ltd, Croydon, CR0 4YY

23/10/2024

01777674-0002